OXFORD MEDICAL PUBLICATIONS

The Psychopharmacology
of Addiction

BRITISH ASSOCIATION
FOR PSYCHOPHARMACOLOGY MONOGRAPHS

The Psychopharmacology of Addiction

BRITISH ASSOCIATION FOR
PSYCHOPHARMACOLOGY MONOGRAPH
No 10
In conjunction with the Society for the Study of Addiction

Edited by

MALCOLM LADER, D.Sc., Ph.D., M.D., F.R.C.Psych.

Professor of Clinical Psychopharmacology
Institute of Psychiatry
University of London

Oxford   New York   Tokyo
OXFORD UNIVERSITY PRESS
1988

Oxford University Press, Walton Street, Oxford OX2 6DP

Oxford New York Toronto
Delhi Bombay Calcutta Madras Karachi
Petaling Jaya Singapore Hong Kong Tokyo
Nairobi Dar es Salaam Cape Town
Melbourne Auckland

and associated companies in
Beirut Berlin Ibadan Nicosia

Oxford is a trade mark of Oxford University Press

Published in the United States
by Oxford University Press, New York

British Library Cataloguing in Publication Data
The Psychopharmacology of addiction
(A British Association for Psychopharmacology
monograph; no. 10) (Oxford medical publications)
1. Drug abuse
I. Lader, M. H.   II. Series
616.86   RC564
ISBN 0–19–261626–9

Library of Congress Cataloging in Publication Data
(Data available)

Typeset by Cotswold Typesetting, Cheltenham
Printed in Great Britain
at the University Printing House, Oxford
by David Stanford
Printer to the University

Preface

The Society for the Study of Addiction was originally founded in 1884 as the Society for the Study and Cure of Inebriety. It held annual meetings and, from the start, published a journal which, after a few changes of title, is now published monthly as the *British Journal of Addiction*.

The British Association for Psychopharmacology was set up in 1974 to encourage the already rapid growth of an important branch of pharmacology and therapeutics. From the start, it concentrated on holding several meetings each year and on its publications, of which this one forms part of the series. Recently, it has set up its own journal, the *Journal of Psychopharmacology*.

I was accorded the honour of being elected President of the SSA in 1983 and of the BAP in 1986. It seemed opportune to combine both my interests, and those of many members of both societies, by holding a scientific meeting devoted to the overlap between the two societies, namely the Psychopharmacology of Addiction. This was held at the Royal Society of Medicine on 24 and 25 November 1986; it was attended by nearly 200 scientists and clinicians covering a most eclectic range of interests. This volume contains the papers given at that meeting. The free communications presented at the meeting are not included.

We were fortunate in having a distinguished group of speakers from the UK, Europe, and North America. In particular, we were gratified that Dr Shepard Siegel from Hamilton, Ontario, delivered the Dent Memorial Lecture.

This volume covers a range of drugs and a variety of techniques and approaches. Thus, the opioids and alcohol are given prominence, but prescribed drugs, such as the benzodiazepines, and licit drugs, such as nicotine and caffeine, are discussed. Techniques range from the complexities of biochemistry to the complexities of biological and social interactions.

I am grateful to the officers of both societies for supporting my initiative in holding this joint symposium, and to the many people who helped in its organization. I am indebted to the speakers for providing their typescripts, and the various staff of Oxford University Press for facilitating the publication.

London
May 1987   M. L.

Contents

Contributors

E. ÄNGGÅRD, *The William Harvey Research Institute,*
St. Bartholomew's Hospital Medical College, Charterhouse Square,
London EC1M 6BQ, UK.

D. BENTON, *Department of Psychology, University College, Swansea,*
SA2 8PP, UK.

S. DOLIN, *Department of Anaesthesiology, Clinical Research Centre,*
Watford Road, Harrow, Middlesex, HA1 3UJ, UK.

R. R. GRIFFITHS, *Department of Psychiatry and Behavioral Sciences,*
and Department of Neuroscience, The Johns Hopkins University School of
Medicine, 720 Rutland Avenue, Baltimore, MD 21205, USA.

J. HARPER, *Department of Pharmacology, King's College, The Strand,*
London, WC2R 2LS, UK.

M. HUDSPITH, *Department of Pharmacology, King's College,*
The Strand, London, WC2R 2LS, UK.

M. LADER, *Department of Psychiatry, Institute of Psychiatry,*
De Crespigny Park, London, SE5 8AF, UK.

H. LITTLE, *Department of Pharmacology, University of Oxford, South*
Parks Road, Oxford, OX1 3QT, UK.

J. LITTLETON, *Department of Pharmacology, King's College,*
The Strand, London, WC2R 2LS, UK.

S. M. MURPHY, *Mapperley Hospital, Porchester Road, Nottingham,*
NG3 6AA, UK.

J. ORFORD, *Department of Psychology, University of Exeter, and*
Department of Clinical Psychology, Exeter Health Authority, Larkby,
Victoria Park Road, Exeter, EX2 4NU, UK.

C. PAGONIS, *Department of Pharmacology, King's College,*
The Strand, London, WC2R 2LS, UK.

S. J. PATERSON, *University of Aberdeen, Unit for Research on*
Addictive Drugs, Marischal College, Aberdeen, AB9 1AS, UK.

S. SIEGEL, *Department of Psychology, McMaster University, Hamilton,*
Ontario, L8S 4K1, Canada.

C. SPYRAKI, *Department of Pharmacology, Medical School,*
University of Athens, Goudi, Athens, 115 27, Greece.

P. TYRER, *Mapperley Hospital, Porchester Road, Nottingham,*
NG3 6AA, UK.

D. M. WARBURTON, *Department of Psychology, University of*
Reading, Building 3, Earley Gate, Whiteknights, Reading, RG6 2AL, UK.

P. P. WOODSON, *Department of Psychiatry and Behavioral Sciences,*
The Johns Hopkins University School of Medicine, 720 Rutland Avenue,
Baltimore, MD 21205, USA.

1

The psychopharmacology of addiction—benzodiazepine tolerance and dependence

MALCOLM LADER

INTRODUCTION

Undoubtedly, psychological and social factors are very important in the institution and maintenance of addictive behaviour. However, it is easy to lose sight of the fact that drugs are involved and that, therefore, pharmacological principles can be applied to study such behaviour. Few pharmacologists have addressed these issues, but of those who have, many of the most distinguished have contributed reviews of the various aspects of the topic to this volume.

This chapter is an attempt to produce a synthesis of pharmacological aspects of tolerance and withdrawal, not of addiction because that is too diffuse a concept. It will seek to relate these concepts to more general properties of psychotropic drugs and, even more widely, to actions of all drugs acting on receptors. The most prescribed class of psychotropic drugs, the benzodiazepines, will be used as an example, but no other generalizations to other substances of concern (such as tobacco, alcohol, cannabis, diamorphine, and cocaine) will be made. The pharmacology of dependence has been dominated too much perhaps by a preoccupation with opioids (Martin 1984). The peptidergic systems in the brain related to opioid action may be a relic of slowly-responding systems unrepresentative of faster-acting neurotransmitter systems which have developed more recently in evolutionary terms.

DEFINITIONS

Tolerance is usually operationally defined as either the reduced effect of the same dose of a drug on subsequent administrations or as a need to increase the dosage in order to maintain the same level of effect. It is generally assumed that the pharmacokinetic, pharmacodynamic, and other mechanisms implicated in the processes of tolerance are the same whether dosage remains the same, and that effect wanes or dosage is increased to maintain the effect. But they might not be so, and there is no evidence either way.

Dependence is a state implicitly assumed to have led to the appearance of behavioural changes (psychological dependence) and/or physiological changes (physical dependence) on withdrawal of the drug. Again, this is a simple operational definition, and it is assumed that the same processes underlie the two types of syndrome.

The term 'drug abuse' is best limited to the excessive non-medical or social use of drugs for their immediate subjective effects, and is akin to other forms of compulsive behaviour such as gambling. The term 'excessive' denotes a value-judgement reflecting social and cultural norms for drug use. The term 'drug addiction' is an unhelpful confusion of these various terms.

Tolerance

The benzodiazepines are undoubtedly effective anxiolytics in the short-term, but the review from the UK Committee on Review of Medicines (1980) concurred with the conclusion of a study carried out by the White House Office of Drug Policy (1979) that benzodiazepines have not been shown to be effective over long periods. The CRM further noted that 'there was little convincing evidence that benzodiazepines were efficacious in the treatment of anxiety after four months' continuous treatment'. One major study since then seems to contradict this view (Rickels *et al.* 1983, 1984, 1985). Chronically anxious out-patients were treated for 6–22 weeks with diazepam (15–40 mg/day), and the efficacy of diazepam was maintained over this period. Withdrawal responses from diazepam can certainly be seen after this period of treatment. If tolerance to the anxiolytic action of diazepam had not developed by this time, then this study provides major evidence against the hypothesis that tolerance is a manifestation of dependence. However, since anxiety levels rarely remain constant, clinical improvement (not drug action) might have been responsible for the decrease in anxiety seen after several weeks of treatment.

In normal volunteers, tolerance to some of the effects of the benzodiazepines seems to develop very rapidly. Tolerance developed to the impairment of driving performance induced by nitrazepam after three doses (Laurell and Tornros 1986). File and Lister (1983) found tolerance to some of the effects of lorazepam, 2.5 mg, after two or three doses, even when the intervals between drug administrations were seven days; the biggest changes were seen from the first to the second dose. The rate of development of tolerance seemed to be system-specific. After three doses it had developed to the decrease in finger-tapping and to self-ratings of dizziness, but not to the drug-induced impairments in learning nonsense-syllable paired-associates, in self-ratings of sedation, or in changes in heart-rate. Ghoneim *et al.* (1981) found that verbal recall was still impaired in normal volunteers after three weeks of daily diazepam administration.

A series of studies on tolerance to benzodiazepines was carried out by Mattila's group in Finland. In one study, volunteers were given lorazepam (1 mg twice a day) or diazepam (5 mg twice a day) for seven days, and evaluated on a battery of tests. No definite tolerance was found on subjective effects, but several psychomotor and cognitive effects showed tolerance to lorazepam, and some cross-tolerance between the two drugs (Aranko *et al.* 1983). Similar studies also failed to demonstrate much tolerance to diazepam (Brosan *et al.* 1986) or to alprazolam (Aranko *et al.* 1985*a*). Cross-tolerance between benzodiazepines is task-dependent, complex tasks being more affected than simple ones (Aranko 1985). In patients taking benzodiazepines for at least one month, dose-dependent development of tolerance to the psychomotor effects of a test dose of lorazepam (3 mg) was demonstrated (Aranko *et al.* 1985*b*).

In long-term benzodiazepine users, the anxiolytic effect, reduction of critical flicker fusion threshold, and the short-term memory impairments induced by benzodiazepines were found to persist, whereas psychomotor impairment or sedation had disappeared (Lucki *et al.* 1986). This suggests that tolerance to the latter effects had developed.

In patients who had been taking normal doses of benzodiazepines for six months or more, tolerance was assessed by giving test doses of diazepam (Petursson and Lader 1984). The responses of patients were compared with the effects of the test dose of diazepam in normal subjects. In the patients, the expected increase in plasma growth hormone concentrations in response to diazepam was almost totally suppressed, indicating marked tolerance. Subjective feelings of sedation to the diazepam were reduced, indicating partial tolerance. There was no tolerance in the EEG fast-wave response. Since the patients were not tested before six months, the onset of tolerance could not be assessed in this study.

Very rapid development of tolerance can be seen after benzodiazepine overdose. In this case, the behavioural effects of benzodiazepines rapidly wane despite persisting high plasma concentrations (Greenblatt *et al.* 1978). This, and the study by Petursson and Lader (1984), suggests that it is possible to demonstrate tolerance after both acute and chronic treatment even when there is still drug acting at the benzodiazepine receptor. This then raises the question of whether withdrawal responses can be elicited when there is still drug acting at the receptor. One animal experiment on withdrawal after acute lorazepam suggests that it is. In the experiment by Lister and Nutt (1987), withdrawal was demonstrated six hours after a single dose of lorazepam, and the withdrawal response was not modified by Ro 15-1788. However, at this time lorazepam was still acting at the receptor, since a residual anticonvulsant action could be detected; this effect was reversed by the receptor antagonist, Ro 15-1788. Most important of all to an analysis of the relative time courses of tolerance and withdrawal, both responses could be seen at the same time after benzodiazepine treatment.

Rebound

Rebound can be defined as the increase in severity of the original symptoms, beyond pre-treatment levels, after short- or long-term drug administration. Rebound effects have been described after sleep laboratory studies involving 1–2 weeks of benzodiazepine administration (Kales *et al.* 1983*a*), as well as after longer-term administration (Adam *et al.* 1976; Oswald *et al.* 1982). Rebound insomnia after the use of benzo-diazepines as hypnotics is now well-documented (for review, see Lader and Lawson 1987). It is more severe than the original insomnia, and is characterized by a delayed onset of sleep and by frequent awakenings. The elimination half-life of the benzodiazepine is important in determining the timing and the severity of rebound. Short-acting drugs (e.g. triazolam, with a half-life about 2.6 hours) produce severe rebound for the next night or two; medium-acting compounds (e.g. temazepam, with a half-life around 8.4 hours) produce less severe rebound two or three nights later; and long-acting drugs (e.g. flurazepam, with a half-life of active metabolites of over 100 hours) produce only minor and sporadic rebound (Kales *et al.* 1983*a*). This phenomenon is quite strongly dose-related; after six consecutive nights of triazolam, normal volunteers showed rebound insomnia after 0.5 mg, but not after 0.25 mg (Roehrs *et al.* 1986). A possible variant of rebound insomnia is early morning insomnia, an increase in wakefulness during the final hours of drug nights. This has been reported following one or two weeks administration of short-acting benzodiazepine hypnotics (Kales *et al.* 1983*b*). Increased anxiety later in the day has also been found (Morgan and Oswald 1982).

Rebound symptoms have also been described following the use for less than six weeks of benzodiazepines to treat anxiety. Pecknold and his colleagues (1982) found rebound anxiety after three weeks of treatment with oxazepam (45 mg/day) or halazepam (120 mg/day). In a placebo-controlled study in general practice, mild rebound anxiety was reported following abrupt termination of treatment after six weeks of 15 mg/day of diazepam (Power *et al.* 1985). The importance of the rate of termination of drug treatment is illustrated by a study in which abrupt termination was compared with a gradual reduction over three weeks, and with placebo administration throughout (Fontaine *et al.* 1984). After four weeks of bromazepam (18 mg/day) or diazepam (15 mg/day), the abruptly-withdrawn patients had ratings on the Hamilton Anxiety scale significantly above their pre-treatment placebo scores, whereas those undergoing gradual drug termination returned to placebo levels. Marked rebound anxiety was more common for the patients stopping bromazepam (5/8) than diazepam (2/8), again illustrating that it is harder to detect rebound effects of benzodiazepines with long half-lives (Busto *et al.* 1986).

Withdrawal

No attempt has been made to define these syndromes with respect to the duration of treatment, but clinically two syndromes have been distinguished on the basis of the dosage involved (Laux and Puryear 1984; Smith and Wesson 1983). However, both may reflect the same underlying mechanism. They are both characterized by symptoms in the direction opposite to the effects of the drug. High-dose (normally 2–5 times the normal anti-anxiety dose) withdrawal has been best characterized by Hollister and his colleagues (1961, 1963), but the literature is peppered with single case reports (Marks 1978; Palmer 1978). Initially, withdrawal after therapeutic dosage was indicated only by sporadic case reports (e.g. Khan *et al.* 1980), but it has now been confirmed in both laboratory and clinical studies (Hallstrom and Lader 1981; Tyrer *et al.* 1981; Petursson and Lader 1981*a*). Even with therapeutic doses there is some evidence that a withdrawal syndrome is found more frequently the longer the treatment, e.g. from similar treatment lasting 22 weeks, rather than six weeks (Rickels *et al.* 1983, 1984). It is difficult to distinguish this syndrome from that described as rebound, except perhaps for category (3) symptoms (below).

Withdrawal symptoms fall roughly into three types (Ladewig (1984);

(1) psychological symptoms of anxiety, such as apprehension, irritability, insomnia, and dysphoria;

(2) bodily symptoms of anxiety, particularly tremor, palpitations, vertigo, sweating, and severe muscle spasms;

(3) perceptual disturbances, such as pains, depersonalization, feeling of motion, metallic taste, and hypersensitivity to light, sound, and touch.

The first two categories may resemble the original anxiety but, as with rebound, the symptoms are more severe (Ladewig 1984). Most commonly these symptoms subside in 5–15 days, which is not consistent with a re-emergence of the original anxiety (Owen and Tyrer 1983). That they are part of a withdrawal response is also indicated by their presence in patients who have been taking benzodiazepines in therepeutic doses for six months or more for a non-psychiatric reason, e.g. chronic muscle spasm following a sports injury (Lader, personal observation).

Gradual withdrawal may be followed by a milder, yet specific, syndrome which is the same whether the dosage was high or low (Hallstrom and Lader 1981). However, even with gradual withdrawal from low doses, prolonged and bizarre responses have been described (Ashton 1984). Ashton emphasizes how physically ill the patients felt, and also describes agoraphobic, panic, and depressive symptoms. In some cases a full-blown depressive syndrome occurs (Olajide and Lader 1984).

In keeping with the evidence that gradual discontinuation of drug treatment gives a less marked withdrawal response than abrupt termination, it has been claimed that short-acting benzodiazepines produce the most marked withdrawal reactions, and that those with elimination half-lives (of parent compound and active metabolites) greater than 36 hours produce milder, but more prolonged, withdrawal (Hollister 1983; Marks 1983). Nonetheless the withdrawal response to drugs in the latter category, e.g. clorazepate and clobazam, is still measurable (Winokur and Rickels 1984; Petursson and Lader 1981b). Support for Hollister's suggestion comes from the findings that diazepam (with its long-acting metabolite, desmethyldiazepam) produces a less severe withdrawal syndrome than the short-acting benzodiazepine, lorazepam (Tyrer *et al.* 1981). A study with high-dose users also suggests that an important factor is the rate of disappearance of drug from the brain. The severity of withdrawal was related to the disappearance of diazepam, and desmethyldiazepam both modified and prolonged the withdrawal syndrome (Rhodes *et al.* 1984). Following abrupt termination of treatment, Tyrer *et al.* (1981) compared the rate of decrease of plasma diazepam and desmethyldiazepam in patients with and without a withdrawal syndrome. Whilst there was no difference for diazepam, more pronounced withdrawal symptoms were accompanied by a more rapid reduction in desmethyldiazepam. When withdrawal took place gradually over four weeks, no such relationship was found, but the rate of decrease of drug concentrations was slower (Tyrer *et al.* 1983).

In conclusion, the duration of action of a particular benzodiazepine might influence both the time at which withdrawal occurs and its severity. The latter is also influenced by the dose and the duration of treatment, but there is no clinical evidence that the withdrawal response is different in nature following different doses. Such differences in metabolic half-life and rate of withdrawal, and their influence on the severity and timing of withdrawal reactions, accord with the factors governing the timing and severity of rebound states. There is thus no evidence for separating either rebound or withdrawal phenomena after low or high doses; they may therefore all be manifestations of the same underlying dependence mechanism.

SELF-ADMINISTRATION AND DRUG PREFERENCE

An extensive series of studies has been carried out by Griffiths and his associates using volunteer human subjects with documented histories of drug abuse. An early experiment established that sodium pento-barbitone, diazepam, and alcohol had similar effects when used as reinforcers; doses of these drugs were exchanged for tokens earned by riding an exercise bicycle (Griffiths *et al.* 1976). A further comparison showed that chlorpromazine (25 or 50 mg/ingestion) was no more effective than placebo as a reinforcer, whereas both pentobarbitone (30 or 90 mg/ingestion) and diazepam (10 or 20 mg/ingestion) maintained self-administration (Griffiths *et al.* 1979). The higher doses were more effective than the lower doses.

In the latter studies from this group, drug preferences were assessed. In addition, rather than repeated moderate doses of the drugs, large single doses were administered. Pentobarbitone, 200–900 mg, was obviously liked by the abuser subjects, and a dose effect was usually clearly present (Griffiths *et al.* 1980). By contrast, diazepam, 50–400 mg, was not clearly associated with subjective liking, nor was any dose-effect obvious. Even observer-rated effects such as ataxia and sedation were poorly related to dose. Diazepam plasma concentrations were, however, commensurate with the dose, suggesting little or no pharmacokinetic tolerance. The preferences between diazepam and pentobarbitone were heavily in favour of the latter.

A direct comparison of the high-potency benzodiazepine, triazolam (0.5–3 mg), and pentobarbitone (100–600 mg) suggested that triazolam had its highest relative potency with relation to objective performance and observer ratings, intermediate potency with subjective ratings of effects, but lowest with subjects' liking and estimate of street value (Roache and Griffiths 1985). It was suggested from these and other data that triazolam has a lower liability to abuse than pentobarbitone, but presents a greater hazard with respect to adverse behavioural effects (Griffiths *et al.* 1985).

Diazepam and oxazepam have also been compared (Griffiths *et al.* 1984). Diazepam (40, 80, and 160 mg) was better liked, produced more euphoria, and was judged to be of greater monetary value than oxazepam (480 mg). Diazepam was only judged a placebo on 4 per cent of trials, oxazepam on 32 per cent. The rapid onset of action of diazepam was deemed a desirable feature for recreational use. The practical validity of such preferences and subjective reports has been explored by examining data for prescription forgeries, theft, and loss in Sweden (Bergman and Griffiths 1986). Prescription rates were similar, but forgeries and loss rates for diazepam were about $2\frac{1}{2}$ times greater than those for oxazepam.

Normal volunteer studies have also been carried out (de Wit *et al.* 1984). These studies are more relevant to drug abuse than to withdrawal problems. They may provide a useful paradigm for the 'spree' use of drugs at weekends within social contexts. Such use often reflects pharmacokinetic factors, such as speed of onset of action, and pharmacodynamic properties such as acute psychotropic, particularly euphoriant, effects (Busto and Sellers 1986).

MECHANISMS OF DEPENDENCE

From the above review, it would seem that the distinction between rebound and withdrawal is a matter of degree rather than of any qualitative difference. Thus, with benzodiazepines, insomnia, anxiety, and bodily symptoms are common to rebound and withdrawal. Although specific withdrawal symptoms have been described (such as perceptual hypersentivity and muscle spasms on benzodiazepine withdrawal) no search for such features has been conducted in rebound using sensitive measures. If we take this step of no longer viewing physical withdrawal syndromes as specific and special to drugs of dependence, then whole areas of both basic and clinical pharmacology become not only relevant but crucial.

The rebound of effect on drug discontinuation is a feature of most drugs, and has been studied with particular relevance to receptor theory. Practical examples include beta-adrenoceptor antagonists and H-2 antagonists, whose withdrawal is known to be followed by an exaggeration of the original symptoms and signs such as hypertension and gastric acidity. Centrally-acting drugs also show rebound hypersensitivity; for example, dopamine-blocking agents such as the antipsychotic drugs may give rebound dyskinesias which are short-lived in contrast to the more severe and prolonged tardive dyskinesias. On discontinuing antidepressants a mixture of peripheral symptoms (such as diarrhoea and salivation) and central symptoms (such as has been noted) occur. Most centrally-acting drugs are followed by some sort of rebound on withdrawal, comprising both somatic and psychological symptoms. Examples include the amphetamines and appetite-suppressants, opioids and other analgesics, ethanol, the barbituates, and the benzodiazepines (see also Haefely 1986 for a discussion of basic mechanisms).

We should, therefore, be looking at the principles which govern the offset of action of all drugs for their reference to the withdrawal of drugs of dependence, rather than singling out these withdrawal syndromes as being in some way special. Certain predictions should follow:

1. All symptoms involved in withdrawal states will show similar to less-marked changes in rebound.

2. All such systems will show changes in the opposite direction during drug onset.

3. Receptor changes following withdrawal may be slow to resolve, thus resulting in prolonged withdrawal.

4. Withdrawal and rebound syndromes with different drugs may not have similar mechanisms unless onset actions affect similar systems.

5. Precise predictions from receptor theory should be possible concerning withdrawal of agonists, antagonists, and mixed agonist/antagonists.

6. Psychological dependence will reflect subjective effects of withdrawal and immediate reinforcing effects.

MECHANISMS OF TOLERANCE

The above review shows very clearly that tolerance is system-specific. Therefore, it can be postulated that different populations of receptors alter at different rates and to different extents when drugs bind to them. But again, receptor changes of this sort are known to accompany the actions of many drugs, both central and peripheral in action. For example, Haefely (1986) has addressed this issue in terms of 'drug-induced adaptive changes'. He also regards tolerance, rebound, and physical dependence as common drug mechanisms not confined to so-called drugs of dependence.

The adaptive changes underlying tolerance presumably vary from system to system and drug to drug, and there is no reason to support the idea of a single common mechanism. Some form of receptor-effector uncoupling is possibly most common, but changes in receptor number and affinity may also occur.

In clinical terms, it is the mechanism which underlies the increase in dose which is relevant. With the benzodiazepines, the bulk of users in the medical context do not escalate their dose but may nevertheless show rebound on withdrawal. Why a few should become clinically tolerant and escalate their dose is unclear. It may even be that such patients increase their dose in order to obtain further psychotropic effects such as sedation or euphoria, and only then does tolerance start to become manifest.

CONCLUSIONS

Tolerance and dependence with drugs of addiction are merely routine examples of such phenomena with drugs in general, but with social consequences. The advantage is that whole areas of classical pharmacology, both animal and clinical, can be combed for their relevance to drug addiction. Leads should become apparent concerning mechanisms involved in dependence problems, and treatments might be devised from simple facts concerning receptors, and screened using *in vitro* preparations.

Nevertheless, other aspects, such as non-social drug abuse, are more complex and may represent special cases relating to the reward properties of only some psychotropic agents. Meanwhile, with many forms of dependence (such as normal-dose benzodiazepine use) pharmacological studies should continue to be worthwhile.

SUMMARY

Tolerance to the benzodiazepines with respect to unwanted effects such as sedation is well-documented. However, it is unclear whether therapeutic effects wane. On discontinuation of benzodiazepines, rebound increases in anxiety, tension, and insomnia are common, the timing and severity being related to the elimination half-life of the compound and to its dosage. More severe rebound merges into withdrawal syndromes in which typical symptom clusters are seen. Abuse of benzodiazepines can be modelled using self-administration and drug preference paradigms. Dependence and withdrawal can be regarded as examples of drug-induced adaptive changes seen with all drug-receptor interactions.

ACKNOWLEDGEMENT

I wish to acknowledge my indebtedness to Dr Sandra File of the Department of Pharmacology, School of Pharmacy, University of London, for many stimulating discussions on the topic. A more detailed discussion of this subject, including animal data, is available (Lader and File, 1987).

REFERENCES

Adam, K., Adamson, L., Brezinova, V., Hunter, W. M., and Oswald, I. (1976). Nitrazepam: lastingly effective but trouble on withdrawal. *Brit. Med. J.* **1**, 1558–60.

Aranko, K. (1985). Task-dependent development of cross-tolerance and motor effects of lorazepam in man. *Acta Pharmacol. Toxicol.* **56**, 373–81.

——, Mattila, M. J. and Seppala, T. (1983). Development of tolerance and cross-tolerance to the psychomotor actions of lorazepam and diazepam in man. *Brit. J. Clin. Pharmacol.* **15**, 545–52.

——, ——, and Bordignon, D. (1985*a*). Psychomotor effects of alprazolam and diazepam during acute and subacute treatment, and during the follow-up phase. *Acta Pharmacol. Toxicol.* **56**, 364–72.

——, ——, Nuutila, A., and Pellinen, J. (1985*b*). Benzodiazepines, but not antidepressants or neuroleptics, induce dose-dependent development of tolerance to lorazepam in psychiatric patients. *Acta Psychiat. Scand.* **72**, 436–46.

Ashton, H. (1984). Benzodiazepine withdrawal: an unfinished story. *Brit. Med. J.* **288**, 1135–40.

Bergman, U. and Griffiths, R. R. (1986). Relative abuse of diazepam and oxazepam: prescription forgeries and theft/loss reports in Sweden. *Drug Alcohol Depend.* **16**, 293–301.

Brosan, L., Broadbent, D., Nutt, D., and Broadbent, M. (1986). Performance effects of diazepam during and after prolonged administration. *Psychol. Med.* **16**, 561–71.

Busto, U. and Sellers, E. M. (1986). Pharmacokinetic determinants of drug abuse and dependence. *Clin. Pharmacokin.* **11**, 144–53.

——, ——, Naranjo, C. A., Cappell, H., Sanchez-Craig, M., and Sykora, K. (1986). Withdrawal reaction after long-term therapeutic use of benzodiazepines. *New Engl. J. Med.* **315**, 854–9.

Committee on the Review of Medicines (1980). Systematic review of the benzodiazepines. *Brit. Med. J.* **1**, 910–12.

De Wit, H., Uhlenhuth, E. H., and Johanson, C. E. (1984). Lack of preference for flurazepam in normal volunteers. *Pharmacol. Biochem. Behav.* **21**, 865–9.

File, S. E. and Lister, R. G. (1983). Does tolerance to lorazepam develop with once weekly dosing? *Brit. J. Clin. Pharmacol.* **16**, 645–50.

Fontaine, R., Chouinard, G., and Annable, L. (1984). Rebound anxiety in anxious patients after abrupt withdrawal of benzodiazepine treatment. *Amer. J. Psychiat.* **141**, 848–52.

Ghoneim, M. M., Mewaldt, S. P., Berie, J. L., and Hinrichs, J. V. (1981). Memory and performance effects of single and 3-week administration of diazepam: 1. Learning and memory. *Psychopharmacol.* **73**, 147–51.

Greenblatt, D. J., Wood, E., Allen, M. D., Orsulak, P. J., and Shader, R. I. (1978). Rapid recovery from massive diazepam overdose. *J. Amer. Med. Assoc.* **240**, 1872–4.

Griffiths, R. R., Bigelow, G. E., and Liebson, I. (1976). Human sedative self-administration: effects of interingestion interval and dose. *J. Pharmacol. Exp. Ther.* **197**, 488–94.

——, ——, and —— (1979). Human drug self-administration: double-blind comparison of pentobarbital, diazepam, chlorpromazine and placebo. *J. Pharmacol. and Exper. Ther.* **210**, 301–10.

——, ——, and —— (1980). Drug preference in humans: double-blind choice comparison of pentobarbital, diazepam and placebo. *J. Pharmacol. Exper. and Ther.* **215**, 649–61.

——, McLeod, D. R., Bigelow, G. E., Liebson, I., and Roache, J. D. (1984). Relative abuse liability of diazepam and oxazepam: behavioural and subjective dose effects. *Psychopharmacol.* **84**, 147–54.

——, Lamb, R. J., Ator, N. A., Roche, J. D., and Bridy, J. V. (1985). Relative abuse liability of triazolam: experimental assessment in animals and humans. *Neurosci. Biobehav. Rev.* **9**, 133–51.

Haefely, W. (1986). Biological basis of drug-induced tolerance, rebound and dependence. Contribution of recent research on benzodiazepines. *Pharmacopsychiat.* **19**, 353–61.

Hallstrom, C. and Lader, M. (1981). Benzodiazepine withdrawal phenomena. *Internat. Pharmacopsychiat.* **16**, 235–44.

Hollister, L. E. (1983). *Clinical pharmacology of psychotherapeutic drugs,* (2nd ed.) p. 41. Churchill Livingstone, New York.

——, Motzenbecker, F. P., and Degan, R. O. (1961). Withdrawal reactions from chlordiazepoxide (Librium). *Psychopharmacologia* **2**, 63–8.

——, Bennett, J. L., Kimbell, I., Jr., Savage, C., and Overall, J. E. (1963). Diazepam in newly admitted schizophrenics. *Dis. Nerv. Sys.* **24**, 746–50.

Kales, A., Soldatos, C. R., Bixler, E. O., and Kales, J. D. (1983*a*). Rebound insomnia and rebound anxiety: a review. *Pharmacol.* **26**, 121–37.

——, ——, ——, and —— (1983*b*). Early morning insomnia with rapidly eliminated benzodiazepines. *Science* **220**, 95–7.

Khan, A., Joyce, P., and Jones, A. V. (1980). Benzodiazepine withdrawal syndromes. *New Zealand Med. J.* **92**, 94–6.

Lader, M. and File, S. (1987). The biological basis of benzodiazepine dependence. *Psychol. Med.* **17**, 539–47.

——, and Lawson, C. (1987). Sleep studies and rebound insomnia: methodological problems, laboratory findings and clinical implications. *Clin. Neuropharmacol.* **10**, 291–312.

Ladewig, D. (1984). Dependence liability of the benzodiazepines. *Drug and Alcohol Dependence* **13**, 139–49.

Laurell, H. and Tornros, J. (1986). The carry-over effects of triazolam compared with nitrazepam and placebo in acute emergency driving situations and in monotonous simulated driving. *Acta Pharmacol. Toxicol.* **58**, 182–6.

Laux, G. and Puryear, D. A. (1984). Benzodiazepines—misuse, abuse and dependency. *Amer. Family Physician* **30**, 139–47.

Lister, R. G. and Nutt, D. J. (in press). Mice and rats are sensitized to the preconvulsant action of a benzodiazepine-receptor inverse agonist (FG 7142) following a single dose of lorazepam. *Brain Res.*

Lucki, I., Rickels, K., and Geller, A. M. (1986). Chronic use of benzodiazepines and psychomotor and cognitive test performance. *Psychopharmacol.* **88**, 426–33.

Marks, J. (1978). *The benzodiazepines—use, overuse, misuse, abuse.* MTP Press, Lancaster.

—— (1983). The benzodiazepines—for good or evil. *Neuropsychobiol.* **10**, 115–26.

Martin, W. R. (1984). Phenomenology and theoretical basis of tolerance and dependence. In *Mechanisms of tolerance and dependence* (ed. C. W. Sharp). NIDA Research Monograph.

Morgan, K. and Oswald, I. (1982). Anxiety caused by a short-life hypnotic. *Brit. Med. J.* **284**, 942.

Olajide, D. and Lader, M. (1984). Depression following withdrawal from long-term benzodiazepine use: a report of four cases. *Psychol. Med.* **14**, 937–40.

Oswald, I., French, C., Adam, K., and Gilham, J. (1982). Benzodiazepine hypnotics remain effective for 24 weeks. *Brit. Med. J.* **284**, 860–3.

Owen, R. T. and Tyrer, P. (1983). Benzodiazepine dependence. A review of the evidence. *Drugs* **25**, 385–98.

Palmer, G. C. (1978). Use, overuse, misuse and abuse of benzodiazepines. *Alabama J. Med. Sci.* **15**, 383–92.

Pecknold., J. C., McClure, D. J., Fleuri, D., and Chang, H. (1982). Benzodiazepine withdrawal effects. *Progr. Neuro-Psychopharmacol. Biol. Psychiat.* **6**, 517–22.

Petursson, H. and Lader, M. (1981*a*). Withdrawal from long-term benzodiazepine treatment. *Br. Med. J.* **283**, 643–5.

—— and —— (1981*b*). Withdrawal symptoms from clobazam. *Royal Soc. Med. Internat. Congr. Symp.* Series No. **43**, 181–3.

—— and —— (1984). *Dependence on tranquillizers.* Maudsley Monograph, No. 28. Oxford University Press, Oxford.

Power, K. G., Jerrom, D. W. A., Simpson, R. J., and Mitchell, M. (1985). Controlled study of withdrawal symptoms and rebound anxiety after six weeks course of diazepam for generalised anxiety. *Brit. Med. J.* **290**, 1246–8.

Rhodes, P. J., Rhodes, R. S., and McCurdy, H. H. (1984). Elimination kinetics and symptomatology of diazepam withdrawal in abusers. *Clin. Toxicol.* **22**, 371–4.

Rickels, K., Case, G. W., Downing, R. W., and Winokur, A. (1983). Long-term diazepam therapy and clinical outcome. *J. Amer. Med. Assoc.* **250**, 767–71.

——, ——, ——, and —— (1985). Indications and contraindications for chronic anxiolytic treatment: is there tolerance to the anxiolytic effect? In *Chronic treatments in neuropsychiatry* (eds. D. Kemali and G. Racagni), pp. 193–204. Raven Press, New York.

——, ——, Winokur, A., and Swenson, C. (1984). Long-term benzodiazepine therapy: benefits and risks. *Psychopharmacol. Bull.* **4**, 608–15.

Roache, J. D. and Griffiths, R. R. (1985). Comparison of triazolam and pentobarbital: performance impairment, subjective effects and abuse liability. *J. Pharmacol. Exp. Ther.* **234**, 120–33.

Roehrs, T. A., Zorick, F. J., Wittig, R. M., and Roth, T. (1986). Dose determinants of rebound insomnia. *Brit. J. Clin. Pharmacol.* **22**, 143–7.

Smith, D. E. and Wesson, D. R. (1983). Benzodiazepine dependence syndromes. *J. Psychoactive Drugs* **15**, 85–95.

Tyrer, P., Rutherford, D., and Huggett, T. (1981). Benzodiazepine withdrawal symptoms and propranolol. *Lancet* **i**, 520–2.

——, Owen, R., and Dawling, S. (1983). Gradual withdrawal of diazepam after long-term therapy. *Lancet* **i**, 1402–6.

White House Office of Drug Policy and National Institute on Drug Abuse (1979). FDA Drug Bulletin, 16 Aug.

Winokur, A. and Rickels, K. (1984). Withdrawal responses to abrupt discontinuation of desmethyldiazepam. *Amer. J. Psychiat.* **141**, 1427–29.

2

The *in vitro* pharmacology of selective opioid ligands

STEWART J. PATERSON

INTRODUCTION

The concept that opioids produce their effects by interacting with different types of receptors was first proposed by Martin (1967) to explain the actions of the potent analgesic, nalorphine (Houde and Wallenstein 1956). Subsequently, Martin and his co-workers carried out a detailed analysis of the effects of a number of opioids in neurophysiological and behavioural tests in dogs, and could distinguish three patterns of activity (Gilbert and Martin 1976; Martin *et al.* 1976). It was proposed that these findings were due to the activation of three different types of opioid receptors, μ, κ, and σ, for which the respective prototype agonists were morphine, ketazocine, and N-allylnormetazocine. However, more recent investigations have led to the suggestion that the σ-effects of opioids are not mediated by opioid receptors but by the phencyclidine receptor (Byrd 1982; Holtzman 1982; Shannon 1982, 1983). Thus, in drug discrimination studies, it was found that animals trained to discriminate between saline and the σ-agonists, N-allylnormetazocine or cyclazocine, responded not only to the opioids, ketazocine and morphine, but also to the non-opioids, phencyclidine and ketamine (Holtzman 1980; Herling *et al.* 1981; Brady *et al.* 1982). This view was supported by the failure of the opioid antagonist naltrexone to block the behavioural effects of either N-allylnormetazocine or phencyclidine in the dog, although it did block the behavioural effects of morphine (Vaupel 1983). Furthermore, the use of stereo-isomers has shown that the (+)-isomers of the opioids were much more potent at producing σ-effects than were the (−)-isomers, whereas the converse was found for the μ- and κ-effects of opioids (Brady *et al.* 1982; Khazan *et al.* 1984; Zukin *et al.* 1984; Mendelsohn *et al.* 1985; Itzhak *et al.* 1985). Thus, there is compelling evidence that the σ-/phencyclidine receptor is a non-opioid receptor.

IN VITRO PHARMACOLOGY

The interpretation of the results obtained in *in vivo* experiments is difficult since the observed effects are influenced by a number of factors, including the distribution and metabolism of the drug after it is administered. The influence of these factors is reduced by the use of isolated tissues in which neurotransmission is sensitive to inhibition by opioids. In the last decade five preparations have been widely used to study the interactions of opioids with their receptors. These are the electrically stimulated guinea-pig ileum (Kosterlitz and Watt 1968), mouse vas deferens (Hughes *et al.* 1975*a*), rat vas deferens (Lemaire *et al.* 1978), rabbit vas deferens (Oka *et al.* 1981) and hamster vas deferens (McKnight *et al.* 1985).

The investigation of the effects of opioids in bioassay preparations has confirmed the existence of μ- and κ-receptors (Hutchinson *et al.* 1975; Chavkin and Goldstein 1981). However, following the discovery of [Met^5]enkephalin and [Leu^5]enkephalin (Hughes *et al.* 1975*b*), the peptides were found to be more active in the mouse vas deferens than in the guinea-pig ileum, whereas the converse was found with the alkaloid, morphine (Lord *et al.* 1977). These findings, with data obtained in binding assays, led to the suggestion that morphine interacts with μ-receptors whereas the enkephalins interact with δ-receptors which are different from μ- or κ-receptors (Lord *et al.* 1977). Therefore, on the basis of the results obtained in bioassay preparations, it would appear that there are three different opioid receptors, μ, δ, and κ. However, results obtained in bioassays using isolated tissues are only unequivocal if the agonists and, more importantly, the antagonists interact exclusively with one type of opioid receptor.

Binding assays

The development of highly selective ligands for the μ-, δ, and κ-receptors has been facilitated by the use of binding assays in homogenates of brain membranes. In this technique, the ability of a ligand to displace the binding of a labelled opioid is a measure of the affinity of the ligand for that binding site. This approach requires the use of labelled ligands which have a high degree of selectivity for each of the three opioid binding sites. The selectivities of some of the available tritiated opioids are shown in Table 2.1. If a ligand interacts with only a single type of opioid site, it would have a relative affinity of 1 at that binding site. It can be seen that the most selective ligands for the three binding sites are [D-Ala^2,MePhe^4,Gly-ol^5]enkephalin (μ), [D-Pen^2,D-Pen^5]enkephalin (δ), and U-69,593 (κ), which all have relative

TABLE 2.1. *The inhibitory effects and relative binding affinity of several opioids available as tritiated ligands for the* μ-, δ- *and* κ-*sites in homogenates of guinea-pig brain at 25°C*

	Inhibitory effect (K_i, nM)			Relative affinity at		
	μ-site	δ-site	κ-site	μ-site	δ-site	κ-site
Morphine	1.80	161	317	0.98	0.01	0.01
[D-Ala²,MePhe⁴,Gly-ol⁵]enkephalin	1.86	407	5590	0.995	0.005	0
[D-Pen²,D-Pen⁵]enkephalin	713	2.56	>15000	0.004	0.996	0
[D-Ser²,L-Leu⁵]enkephalyl-Thr	35.6	1.78	6041	0.05	0.95	0
[D-Ala²,D-Leu⁵]enkephalin	12.0	1.35	12400	0.10	0.90	0
(−)Ethylketazocine	1.01	6.62	1.02	0.47	0.07	0.46
(−)Bremazocine	0.62	0.78	0.15	0.17	0.13	0.70
Dynorphin A (1-9)	3.30	3.17	0.228	0.06	0.06	0.88
U-69,593*	2350	19670	5.58	0.003	0	0.997

The μ-binding sites were labelled with [³H]-[D-Ala²,MePhe⁴,Gly-ol⁵]enkephalin (1 nM).
The δ-binding sites were labelled with [³H]-[D-Pen²,D-Pen⁵]enkephalin (1.6 nM) or [³H]-[D-Ala²,D-Leu⁵]enkephalin (0.7 nM) in the presence of 30 nM unlabelled [D-Ala²,MePhe⁴,Gly-ol⁵]enkephalin.
The κ-binding sites were labelled with [³H]-(−)-bremazocine (0.1–0.3 nM) in the presence of 100 nM unlabelled [D-Ala²,MePhe⁴,Gly-ol⁵]enkephalin and 100 nM unlabelled [D-Ala²,D-Leu⁵]enkephalin.
The binding of dynorphin A (1-9) was determined at 0°C.
Relative affinity is (K_i at μ, δ or κ)$^{-1}$/(K_i^{-1} at μ+K_i^{-1} at δ+K_i^{-1} at κ).
Pen is penicillamine.
*U-69,593 is (5α,7α,8β)-(+)-N-methyl-N-(7-(1-pyrrolidinyl)-1-oxaspiro[4,5]dec-8-yl)benzeneacetamide.

affinities of > 0.995. If it is not possible to use these selective tritiated compounds, the individual sites can be labelled by restricting the binding of a non-selective or less-selective labelled ligand to a single site by preventing binding at the other sites with the appropriate unlabelled ligands (Gillan and Kosterlitz 1982; Corbett *et al.* 1984; Gillan *et al.* 1985; Robson *et al.* 1985). Thus, the binding of [³H]-bremazocine or [³H]-dynorphin A (1–9) can be restricted to the κ-site by the addition of the unlabelled μ-ligand [D-Ala², MePhe⁴, Gly-ol⁵]enkephalin and the unlabelled δ-ligand [D-Ala², D-Leu⁵]enkephalin. Similarly, the binding of non-selective δ-ligands, e.g. [D-Ala², D-Leu⁵]enkephalin, can be limited to the δ-sites by addition of the unlabelled μ-ligand [D-Ala², MePhe⁴, Gly-ol⁵]enkephalin.

When the opioid binding sites in the brain are selectively labelled, it is found that the proportions of the three sites are different in different species (Table 2.2.). In the rat, rabbit, and mouse μ-sites are the most numerous, whereas in the guinea-pig κ-sites predominate. However, the proportions of the three sites vary independently within the species. Similar differences are found in the regional distribution of the μ-, δ-, and κ-binding sites (Chang *et al.* 1981; Ninkovic *et al.* 1981; Pfeiffer *et al.* 1982; Robson *et al.* 1985). One region of particular interest is the cerebellum. Although the rat cerebellum is devoid of opioid binding sites (Pert and Snyder 1973; Meunier and Zajac 1979), the guinea-pig cerebellum contains sites which are mainly of the κ-type (Robson *et al.* 1984) whereas the rabbit cerebellum contains mainly μ-sites (Meunier 1982; Frances *et al.* 1985). Therefore these tissues can be used to restrict the binding of a tritiated ligand largely to the κ- or μ-sites respectively.

TABLE 2.2. *Species distribution of opioid binding sites*

Species	Total opioid binding capacity (pmol/g brain)	Proportion of sites		
		μ	δ	κ
rat	15.8	46	42	12
rabbit	12.0	43	19	38
DBA/2 mouse	14.4	51	29	20
guinea-pig	12.3	24	32	44

Data from Robson *et al.* (1985).

The selectivity of a number of opioids has been assessed in homogenates of guinea-pig brain using selective labelling techniques (Magnan *et al.* 1982; Corbett *et al.* 1984; Cotton *et al.* 1985). The most selective agonists and antagonists for each of the three sites are shown in

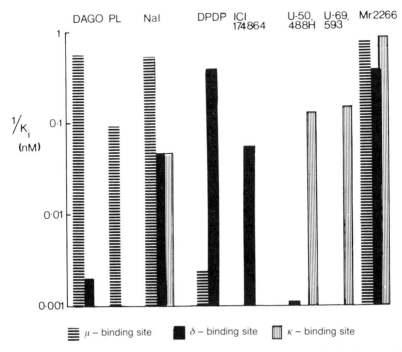

FIG. 2.1.   The affinities for the μ-, δ- and κ-binding sites of selective agonists and antagonists at each receptor. The μ-, δ- and κ-binding sites were labelled as described in Table 2.1.
DAGO: [D-Ala^2,MePhe^4,Gly-ol^5]enkephalin;   PL: Tyr-Pro-MePhe-D-Pro-NH$_2$;
NAL: naloxone;   DPDP: [D-Pen^2,D-Pen^5]enkephalin;
ICI 174864: N,N-diallyl-Tyr-Aib-Aib-Phe-Leu (Aib = α-aminoisobutyric acid);
U-50,488H: trans-3,4-dichloro-N-methyl-N-(2-(1-pyrrolidinyl)-cyclohexyl)-
benzeneacetamide;   U-69,593: (5α,7α,8β)-(+)-N-methyl-N-(7-(1-pyrrolidinyl)-
1-oxaspiro[4,5]dec-8-yl)-benzeneacetamide;   Mr 2266: ($-$)-α-5,9-diethyl-2-
(3-furyl-methyl)-2'-hydroxy-6,7-benzomorphan.

Figure 2.1. The most selective ligands for the μ-binding site are Tyr-Pro-MePhe-D-Pro-NH$_2$ and [D-Ala^2,MePhe^4,Gly-ol^5]enkephalin. On the other hand, the μ-antagonist naloxone is considerably less selective for the μ-site than is either of the agonists. At the δ-binding site [D-Pen^2,D-Pen^5]enkephalin and ICI 174864 are both highly selective ligands but the agonist [D-Pen^2,D-Pen^5]enkephalin is considerably more potent than the antagonist ICI 174864. At the κ-binding site, the agonists U-50,488H and U-69,593 are the most selective ligands, whereas the antagonist Mr 2266 is only slightly more active at the κ-site than the μ-site.

Pharmacological assays

The effects of agonists

The unequivocal characterization of the receptors involved in mediating the pharmacological actions of opioids can only be achieved with highly selective ligands. Since binding assays do not provide information about whether a compound interacts with a receptor to produce an agonist response or is an antagonist, it is necessary to compare the results obtained in the binding assays with those in pharmacological assays.

The selective μ-ligands Tyr-Pro-MePhe-D-Pro-NH$_2$ and [D-Ala^2,MePhe^4,Gly-ol^5]enkephalin are potent agonists in both the guinea-pig ileum and the mouse vas deferens (Handa *et al.* 1981; Chang *et al.* 1983; Corbett *et al.* 1984). [D-Ala^2,MePhe^4,Gly-ol^5]enkephalin is also an agonist in the rat vas deferens (Gillan *et al.* 1981) but has no agonist action in the hamster vas deferens (Corbett *et al.* 1984; McKnight *et al.* 1985).

The selective δ-ligand [D-Pen^2,D-Pen^5]enkephalin is a potent agonist in the hamster vas deferens and in the mouse vas deferens (Mosberg *et al.* 1983; Corbett *et al.* 1984). On the other hand [D-Pen^2,D-Pen^5]enkephalin has no agonist activity in the vasa deferentia of the rat and rabbit (Corbett *et al.* 1984). In the guinea-pig ileum, [D-Pen^2,D-Pen^5]enkephalin is a weak agonist with an IC$_{50}$ value greater than 3000 nM (Mosberg *et al.* 1983; Corbett *et al.* 1984). However, this effect is mediated by μ-receptors since the K$_e$ value for antagonism by naloxone is 2.7 nM (Mosberg *et al.* 1983).

The selective κ-compounds, U-50,488H and U-69,593, are potent agonists in the guinea-pig ileum and in the rabbit vas deferens (Gillan *et al.* 1983; Corbet *et al.* 1985a; Hayes and Kelly 1985; Hayes *et al.* 1985) but are inactive in the vasa deferentia of the hamster and the rat (Gillan *et al.* 1983; Corbett *et al.* 1985a; McKnight *et al.* 1985). U-50,488H is also a potent agonist in the mouse vas deferens (Romer *et al.* 1984).

These findings with the agonists which were highly selective ligands in the binding assays confirm that the hamster vas deferens contains only δ-receptors (McKnight *et al.* 1985), the rabbit vas deferens contains only κ-receptors (Oka *et al.* 1981), the guinea-pig ileum contains μ- and κ-receptors (Hutchinson *et al.* 1975; Chavkin and Goldstein 1981) and the mouse vas deferens contains all three receptors (Lord *et al.* 1977). Since in the rat vas deferens the selective μ-ligand [D-Ala^2,MePhe^4,Gly-ol^5]enkephalin is an agonist but selective δ- and κ-ligands are inactive, this preparation appears to contain μ-receptors (Gillan *et al.* 1981). This finding does not preclude the possibility that it also contains ε-receptors which display a high affinity for β-endorphin (Schulz *et al.* 1979).

The effects of antagonists

The lack of selectivity of the μ-antagonist naloxone is highlighted by the fact that it is an antagonist in all five bioassay preparations. In the guinea-pig ileum, naloxone readily antagonises the μ-agonists Tyr-Pro-MePhe-D-Pro-NH$_2$ and [D-Ala^2,MePhe^4,Gly-ol^5]enkephalin with K$_e$ values less than 3 nM (Handa *et al.* 1981; Chang *et al.* 1983). In contrast, naloxone is a less effective antagonist of the κ-agonist U-69,593 with a K$_e$ value of 18 nM (Corbett *et al.* 1985*a*). Similarly, in the mouse vas deferens, naloxone antagonises the μ-agonists Tyr-Pro-MePhe-D-Pro-NH$_2$ and . [D-Ala^2,MePhe^4,Gly-ol^5]enkephalin with K$_e$ values less than 3 nM (Handa *et al.* 1981; Gillan *et al.* 1981; Chang *et al.* 1983) whereas it is less effective against the δ-agonist [D-Pen^2,D-Pen^5]enkephalin with a K$_e$ value of 46 nM (Mosberg *et al.* 1983). In the hamster vas deferens, naloxone is an antagonist of the selective δ-agonist [D-Pen^2,D-Pen^5]enkephalin with a K$_e$ value of 51 nM (McKnight *et al.* 1985). In addition, in the rabbit vas deferens, naloxone antagonises the inhibitory effects of the κ-agonist U-50,488H with a K$_e$ value of 25 nM (Gillan *et al.* 1983). Thus, as in the binding assays, naloxone has highest affinity for μ-receptors and is less active at both δ- and κ-receptors.

The selective δ-ligand ICI 174864 is an antagonist of the selective δ-agonist [D-Pen^2,D-Pen^5]enkephalin in the hamster vas deferens (McKnight *et al.* 1985; Miller and Shaw 1985); the K$_e$ values vary between 30 nM and 51 nM. The high degree of δ-selectivity found for ICI 174864 in the binding assays is confirmed by the observations that it has no antagonist action at 5000 nM against μ- or κ-agonists in the mouse vas deferens (Cotton *et al.* 1984). Similarly, in the guinea-pig ileum, 2000 nM ICI 174864 does not antagonise the effects of the μ-agonist [D-Ala^2,MePhe^4,Gly-ol^5]enkephalin (Takemori and Portoghese 1985).

Since Mr 2266 is not a selective ligand in the binding assays, it is not unexpected that it is an antagonist at all three receptors. In the rabbit vas deferens, which contains only κ-receptors, Mr 2266 is a more potent antagonist than naloxone (Oka *et al.* 1982; Gillan *et al.* 1983). In the guinea-pig ileum, which contains both μ- and κ-receptors, Mr 2266 is a potent antagonist of the μ-agonist normorphine and the κ-agonist dynorphin A (Lord *et al.* 1977; Yoshimura *et al.* 1982). Similarly, in the mouse vas deferens (which contains all three receptors) Mr 2266 is a potent antagonist of normorphine and dynorphin A, but is a less effective antagonist of the δ-agonists, [Met^5]enkephalin and [Leu^5]enkephalin (Lord *et al.* 1977; James *et al.* 1984).

The results obtained with the antagonists in the pharmacological assays confirm that the δ-ligand ICI 174864 is the only highly selective antagonist available at present. Although naloxone has a preference for the μ-receptor, more selective μ-antagonists and selective κ-antagonists are urgently required for the investigation of the physiological role of opioids.

The use of selective assay tissues

Although the hamster vas deferens is a selective δ-bioassay tissue and the rabbit vas deferens is a selective κ-bioassay tissue, there is no selective μ-bioassay tissue. However, the contractions of the guinea-pig ileum are very sensitive to inhibition by both κ- and μ-agonists. This tissue can be used as a selective κ- or μ-bioassay tissue if it is pre-treated with the opioid alkylating agents β-funaltrexamine or β-chlornaltrexamine (Portoghese *et al.* 1979, 1980; Goldstein and James 1984). β-funaltrexamine preferentially alkylates μ-receptors whereas β-chlornaltrexamine alkylates both μ- and κ-receptors. After exposure of the tissue to either agent, the residual alkylating agent can be removed by extensive washing of the tissue. After pre-treatment with β-funaltrexamine, the agonist potency of the μ-ligand [D-Ala^2,MePhe^4,Gly-ol^5]enkephalin is reduced, whereas the agonist potencies of the κ-agonists U-50,488H and U-69,593 are not altered (Corbett *et al.* 1985*b*; Corbett and Kosterlitz 1986). In contrast, exposure of the guinea-pig ileum to β-chlornaltrexamine reduces the agonist potencies of both [D-Ala^2,MePhe^4,Gly-ol^5]enkephalin and U-69,593 (Corbett and Kosterlitz 1986). However, if the μ-ligand [D-Ala^2,MePhe^4,Gly-ol^5]enkephalin is present during the exposure to β-chlornaltrexamine, the μ-receptors can be selectively protected from alkylation. After this pre-treatment, the IC$_{50}$ value for [D-Ala^2,MePhe^4,Gly-ol^5]enkephalin is increased two-fold, whereas that of U-69,593 is increased 32-fold. Therefore, the tissue can now be used as a μ-selective bioassay tissue (Corbett and Kosterlitz 1986).

CONCLUSION

In conclusion, it is now possible to label the μ-, δ- and κ-binding sites either with selective ligands or by the use of techniques which restrict the binding of the tritiated ligand to a single site. In addition, it is possible to selectively study the pharmacology of the three receptors in the hamster vas deferens (δ), the rabbit vas deferens or the guinea-pig ileum after pre-treatment with β-funaltrexamine (κ) and the guinea-pig ileum after pre-treatment with β-chlornaltrexamine in the presence of [D-Ala^2,MePhe^4,Gly-ol^5]enkephalin (μ). The use of these techniques will facilitate the development of the more selective opioid agonists and antagonists which are essential for a fuller understanding of the physiological role of opioids in man.

ACKNOWLEDGEMENTS

Supported by grants from the Medical Research Council and the US National Institute on Drug Abuse (DA 00662).

REFERENCES

Brady, K. T., Balster, R. L., and May, E. L. (1982). Stereoisomers of N-allylnormetazocine: phencyclidine-like behavioral effects in squirrel monkeys and rats. *Science* **215**, 178–80.

Byrd, L. D. (1982). Comparison of the behavioral effects of phencyclidine, ketamine, d-amphetamine and morphine in the squirrel monkey. *J. Pharm. and Exper. Ther.* **220**, 139–44.

Chang, K.-J., Hazum, E., and Cuatrecasas, P. (1981). Novel opiate binding sites selective for benzomorphan drugs. *Proc. Nat. Academy of Sci. of the USA* **78**, 4141–5.

——, Wei, E. T., Killian, A., and Chang, J. K. (1983). Potent morphiceptin analogs: structure activity relationships and morphine like activities. *J. Pharm. and Exper. Ther.* **227**, 403–8.

Chavkin, C. and Goldstein, A. (1981). Demonstration of a specific dynorphin receptor in guinea-pig ileum myenteric plexus. *Nature (London)* **29**, 591–3.

Corbett, A. D. and Kosterlitz, H. W. (1986). Bremazocine is an agonist at κ-opioid receptors and an antagonist at μ-opioid receptors in the guinea-pig myenteric plexus. *Brit. J. Pharm.* **89**, 245–9.

——, Gillan, M. G. C., Kosterlitz, H. W., McKnight, A. T., Paterson, S. J., and Robson, L. E. (1984). Selectivities of opioid peptide analogues as agonists and antagonists at the δ-receptor. *Brit. J. Pharm.* **83**, 271–9.

——, ——, ——, and Paterson, S. J. (1985*a*). Binding and pharmacological profile of a highly selective ligand for the κ-opioid receptor—U-69,593. *Brit. J. Pharm.* **86**, 704P.

——, Kosterlitz, H. W., McKnight, A. T., Paterson, S. J., and Robson, L. E. (1985*b*). Pre-incubation of guinea-pig myenteric plexus with β-funaltrexamine: discrepancy between binding assays and bioassays. *Brit. J. Pharm.* **85**, 665–73.

Cotton, R., Giles, M. G., Miller, L., Shaw, J. S., and Timms, D. (1984). ICI 174864: A highly selective antagonist of the δ-opioid receptor. *European J. Pharm.* **97**, 331–2.

——, Kosterlitz, H. W., Paterson, S. J., Rance, M. J., and Traynor, J. R. (1985). The use of [^{3}H]-[D-Pen^2,D-Pen^5]enkephalin as a highly selective ligand for the δ-binding site. *Brit. J. Pharm.* **84**, 927–32.

Frances, B., Moisand, C., and Meunier. J.-C. (1985). Na$^+$ ions and Gpp(NH)p selectively inhibit agonist interactions at μ- and κ-opioid receptor sites in rabbit and guinea-pig cerebellum membranes. *European J. Pharm.* **117**, 223–32.

Gilbert, P. E. and Martin, W. R. (1976). The effects of morphine- and nalorphine-like drugs in the non-dependent, morphine-dependent and cyclazocine-dependent chronic spinal dog. *J. Pharm. and Exper. Ther.* **198**, 66–82.

Gillan, M. G. C. and Kosterlitz, H. W. (1982). Spectrum of the μ-, δ- and κ-binding sites in homogenates of rat brain. *Brit. J. Pharm.* **77**, 461–9.

——, ——, and Magnan, J. (1981). Unexpected antagonism in the rat vas deferens by benzomorphans which are agonists in other pharmacological tests. *Brit. J. Pharm.* **72**, 13–15.

——, Jin, W-Q., Kosterlitz, H. W., and Paterson, S. J. (1983). A highly selective ligand for the κ-binding site (U-50,488H). *Brit. J. Pharm.* **79**, 275P.

——, Robson, L. E., McKnight, A. T., and Kosterlitz, H. W. (1985). κ-Binding and degradation of [^{3}H]dynorphin A (1–8) and [^{3}H]dynorphin A (1–9) in suspensions of guinea pig brain membranes. *J. Neurochem.* **45**, 1034–42.

Goldstein, A. and James, I. F. (1984). Site-directed alkylation of multiple opioid receptors. 11. Pharmacological selectivity. *Molecular Pharm.* **25**, 343–8.

Handa, B. K., Lane, A. C., Lord, J. A. H., Morgan, B. A., Rance, M. J., and Smith, C. F. C. (1981). Analogues of β-LPH61-64 possessing selective agonist activity at μ-opiate receptors. *European J. Pharm.* **70**, 531–40.

Hayes, A. and Kelly, A. (1985). Profile of activity of κ-receptor agonists in the rabbit vas deferens. *European J. Pharm.* **110**, 317–22.

——, Sheehan, M. J., and Tyers, M. B. (1985). Determination of the receptor selectivity of opioid agonists in the guinea-pig ileum and mouse vas deferens by use of β-funaltrexamine. *Brit. J. Pharm.* **86**, 899–904.

Herling, S., Coale, E. H., Hein, D. W., Winger, G., and Woods, J. H. (1981). Similarity of the discriminative stimulus effects of ketamine, cyclazocine and dextrorphan in the pigeon. *Psychopharmacology* **73**, 286–91.

Holtzman, S. G. (1980). Phencyclidine-like discriminative effects of opioids in the rat. *J. Pharm. and Exper. Ther.* **214**, 614–19.

—— (1982). Phencyclidine-like discriminative stimulus properties of opioids in the squirrel monkey. *Psychopharmacology* **77**, 295–300.

Houde, R. W. and Wallenstein, S. L. (1956). Clinical studies of morphine-nalorphine combinations. *Federation Proc.* **15**, 440–1.

Hughes, J., Kosterlitz, H. W., and Leslie, F. M. (1975a). Effect of morphine on adrenergic transmission in the mouse vas deferens. Assessment of agonist and antagonist potencies of narcotic analgesics. *Brit. J. Pharm.* **53**, 371–81.

——, Smith, T. W., Kosterlitz, H. W., Fothergill, L. A., Morgan, B. A., and Morris, H. R. (1975b). Identification of two related pentapeptides from the brain with potent opiate agonist activity. *Nature (London)* **258**, 577–9.

Hutchinson, M., Kosterlitz, H. W., Leslie, F. M., Waterfield, A. A., and Terenius, L. (1975). Assessment in the guinea-pig ileum and mouse vas deferens of benzomorphans which have strong antinociceptive activity but do not substitute for morphine in the dependent monkey. *Brit. J. Pharm.* **55**, 541–6.

Itzhak, Y., Hiller, J. M., and Simon, E. J. (1985). Characterisation of specific binding sites for [^{3}H](d)-N-allylnormetazocine in rat brain membranes. *Molecular Pharm.* **27**, 46–52.

James, I. F., Fischli, W., and Goldstein, A. (1984). Opioid receptor selectivity of dynorphin gene products. *J. Pharm. and Exper. Ther.* **228**, 88–93.

Khazan, N., Young, G. A., El-Fakary, E. S., Hong, O., and Calligaro, D. (1984). Sigma receptors mediate the psychotomimetic effects of N-allylnormetazocine

(SKF-10,047), but not its opioid agonistic-antagonistic properties. *Neuro-pharmacology* **23**, 983–7.

Kosterlitz, H. W. and Watt, A. J. (1968). Kinetic parameters of narcotic agonists and antagonists, with particular reference to N-allylnoroxy-morphone (naloxone). *Brit. J. Pharm.* **33**, 266–76.

Lemaire, S., Magnan, J., and Regoli, D. (1978). Rat vas deferens: a specific bioassay for endogenous opioid peptides. *Brit. J. Pharm.* **64**, 327–9.

Lord, J. A. H., Waterfield, A. A., Hughes, J., and Kosterlitz, H. W. (1977). Endogenous opioid peptides: multiple agonists and receptors. *Nature (London)* **267**, 495–9.

McKnight, A. T., Corbett, A. D., Marcoli, M., and Kosterlitz, H. W. (1985). The opioid receptors in the hamster vas deferens are of the δ-type. *Neuro-pharmacology* **24**, 1011–7.

Magnan, J., Paterson, S. J., Tavani, A., and Kosterlitz, H. W. (1982). The binding spectrum of narcotic analgesic drugs with different agonist and antagonist properties. *Naunyn-Schmiedeberg's Archives of Pharmacology* **319**, 197–205.

Martin, W. R. (1967). Opioid antagonists. *Pharmacological Reviews* **19**, 463–521.

——, Eades, C. G., Thompson, J. A., Huppler, R. E., and Gilbert, P. E. (1976). The effects of morphine- and nalorphine-like drugs in the nondependent and morphine-dependent chronic spinal dog. *J. Pharm. and Exper. Ther.* **197**, 517–32.

Mendelsohn, L. G., Kalra, V., Johnson, B. G., and Kerchner, G. A. (1985). Sigma opioid receptor: characterisation and co-identity with the phencyclidine receptor. *J. Pharm. and Exper. Ther.* **233**, 597–608.

Meunier, J.-C. (1982). Mu and kappa opiate binding sites in the rabbit CNS. *Life Sciences* **31**, 1327–30.

—— and Zajac, J.-M. (1979). Cerebellar opiate receptors in lagomorphs. Demonstration, characterisation and regional distribution. *Brain Research* **168**, 311–21.

Miller, L. and Shaw, J. S. (1985). Characterisation of the δ-opioid receptor on the hamster vas deferens. *Neuropeptides* **6**, 531–6.

Mosberg, H. I., Hurst, R., Hruby, V. J., Gee, K., Yamamura, H. I., Galligan, J. J., and Burks, T. F. (1983). Bis-penicillamine enkephalins possess highly improved specificity toward δ-opioid receptors. *Proc. Nat. Acad. of Sci. of the USA* **80**, 5871–4.

Ninkovic, M., Hunt, S. P., Emson, P. C., and Iversen, L. L. (1981). The distribution of multiple opiate receptors in bovine brain. *Brain Research* **214**, 163–7.

Oka, T., Negishi, K., Suda, M., Matsumiya, T., and Ueki, M. (1981). Rabbit vas deferens: a specific bioassay for opioid κ-receptor agonists. *European J. Pharm.* **73**, 235–6.

——, ——, ——, Sawa, A., Fujino, M., and Wahimasu, M. (1982). Evidence that dynorphin (1–13) acts as an agonist on opioid κ-receptors. *European J. Pharm.* **77**, 137–41.

Pert, C. B. and Snyder, S. H. (1973). Opiate receptor: demonstration in nervous tissue. *Science* **179**, 1011–14.

Pfeiffer, A., Pasi, A., Mehraein, P., and Herz, A. (1982). Opiate receptor binding sites in human brain. *Brain Research* **248**, 87–96.

Portoghese, P. S., Larson, D. L., Jiang, J. B., Caruso, T. P., and Takemori, A. E. (1979). Synthesis and pharmacological characterization of an alkylating analogue (chlornaltrexamine) of naltrexone with ultralong-lasting narcotic antagonist properties. *J. Medicinal Chemistry* **22**, 168–73.
——, ——, Sayre, L. M., Fries, D. S., and Takemori, A. E. (1980). A novel opioid receptor site directed alkylating agent with irreversible narcotic antagonistic and reversible agonistic activities. *J. Medicinal Chemistry* **26**, 1341–3.
Robson, L. E., Foote, R. W., Maurer, R., and Kosterlitz, H. W. (1984). Opioid binding sites of the κ-type in guinea-pig cerebellum. *Neuroscience* **12**, 621–7.
——, Gillan, M. G. C., and Kosterlitz, H. W. (1985). Species differences in the concentrations and distributions of opioid binding sites. *European J. Pharm.* **112**, 65–71.
Romer, D., Maurer, R., and Hill, R. C. (·1984). Kappa agonists: pharmacological characterisation. In *Quo vadis? Analgesia and enkephalinases. Kappa receptors and their ligands* (eds. R. Biogegrain, J. Cros, M. Morre, J. P. Meyard, and R. Roncucci) pp. 450–62. Sanofi Research, Montpellier.
Schulz, R., Faase, E., Wuster, M., and Herz, A. (1979). Selective receptors for β-endorphin on the rat vas deferens. *Life Sciences* **24**, 843–9.
Shannon, H. E. (1982). Phencyclidine-like discriminative stimuli of (+) and (−)-N-allylnormetazocine in rats. *European J. Pharm.* **84**, 225–8.
—— (1983). Pharmacological evaluation of N-allylnormetazocine (SKF 10,047) on the basis of its discriminative stimulus properties in the rat. *J. Pharm. and Exper. Ther.* **225**, 144–8.
Takemori, A. E. and Portoghese, P. S. (1985). Receptors for opioid peptides in the guinea-pig ileum. *J. Pharm. and Exper. Ther.* **235**, 389–92.
Vaupel, D. B. (1983). Naltrexone fails to antagonise the σ-effects of PCP and SKF 10,047 in the dog. *European J. Pharm.* **92**, 269–74.
Yoshimura, K., Huidobro-Toro, J. P., and Way, E. L. (1982). Potency of three opiate antagonists to reverse the inhibitory activity of dynorphin, enkephalin and opioid-like alkaloids on the guinea-pig ileum. *European J. Pharm.* **84**, 17–24.
Zukin, S. R., Brady, K. T., Slifer, B. L., and Balster, R. L. (1984). Behavioural and biochemical stereoselectivity of sigma opiate/PCP receptors. *Brain Research* **294**, 174–7.

3

The puzzle of nicotine use

DAVID M. WARBURTON

INTRODUCTION

There is now sufficient evidence supporting the hypothesis that nicotine is the major chemical agent underlying the smoking habit. Certainly it would be a most remarkable coincidence that most smokers restrict their smoking habits to a leaf which is the only one which also contained nicotine. As a result of inhalation, a 20-cigarette per day smoker will receive over 70 000 boluses of nicotine per year. It has been estimated that, with inhalation, this drug reaches the brain in seven seconds, over twice as fast as the drug heroin when main-lined into the arm. These 70 000 rapid deliveries of a drug to the brain cause marked changes in brain activity. As Jarvik (1973) remarked, 'It would be a remarkable coincidence if the effects of this powerful pharmacological agent had nothing to do with why people smoke'.

The puzzle of nicotine use concerns the mechanisms of action that account for its use. As will become clear as this chapter proceeds, the psychological effects of nicotine are much more subtle than those of other drugs, such as alcohol or heroin, and thus particularly hard to quantify. Nonetheless, evidence will be presented indicating that the psychological effects of nicotine are crucial to the maintenance of the smoking habit in the vast majority of smokers, although the mechanisms by which the drug achieves this effect are quite different from those underlying other habitual substance use.

EUPHORIANT MODEL

Most other habitually-used substances are positively reinforcing because they have psychological effects which are described as pleasurable. In traditional accounts of substance use, it has been common to explain their use in terms of a state of 'euphoria'; this is not simply a reduction of

27

physiological or psychological pain, but a positive experience of feeling right with oneself, and with the world (Blum 1984). This euphoria is the outcome of the compounds acting on the same underlying systems in the brain which are mediating pleasure for normal human experiences. If nicotine were producing such effects by acting upon similar brain mechanisms then we would expect a number of similarities in the mode of action of the drug with the action of other experiences.

Activity in the pleasure pathways

In the Euphoriant Model, drugs come to dominate motivated behaviour because they activate the endogenous motivational pathways of the brain, and thus serve as hedonically effective substitutes for biologically important rewards. The view that there are specialized neural systems subserving the rewarding effects of such natural rewards as food, water, and sex has been assumed by many since the demonstration that animals would work to stimulate some of their brain pathways electrically. The dominant theories of brain-stimulation reward propose that there is activation of catecholaminergic fibres for rewarding brain stimulation (Warburton 1975). These pathways are part of the medial forebrain bundle, brain-stimulation reward system which appears to be activated synaptically by the sensory inputs which carry the signals of food and water rewards (Wise 1982).

The major reason for believing that habitually-used drugs act on the brain 'pleasure' system is that these drugs facilitate brain-stimulation reward. Alcohol (St Laurent and Olds 1967), amphetamine (Stein 1962), barbiturates (Mogenson 1964), cocaine (Crow 1970), heroin (Koob *et al.* 1975), and morphine (Lorens and Mitchell 1973) have been reported as facilitating brain-stimulation responding.

Given that these agents share the capacity for activating a common reward system, the Euphoriant Model argues that nicotine would increase self-stimulation responding in animal studies. However, the drug had little effect upon self-stimulation at doses that approximate human smoking doses (i.e. 0.1 mg/kg to 0.4 mg/kg in the rat). It has even been suggested that the drug decreases self-stimulation at doses that approximate human smoking (Domino 1973).

Neurochemical mode of action

The second expectation from the Euphoriant Model is that nicotine would act on the neurochemical systems which are believed to control pleasure, the catecholamine [1] and endogenous opiate systems. The possibility of an action on catecholamine systems has been poorly supported. Although there is some evidence that nicotine affects noradrenalin and dopamine, this effect is secondary to the action of the drug upon cholinergic neurones in the midbrain and only occurs at very high doses (Fuxe *et al.* 1977). More recently, the endogenous opiate hypothesis has gained prominence.

Specific opiate binding sites have been found in the brain, and this finding has been followed by the discovery of a number of endogenous peptide molecules which mimic morphine. The most interesting from our point of view is beta-endorphin. Beta-endorphin mimics morphine not only after a single injection, but also after repeated injections. Repeated dosing with beta-endorphin leads to development of tolerance and physical dependence (Van Ree *et al.* 1979). Brain endorphins seem to be involved in brain-stimulation-reward, since the opiod antagonist, naloxone, suppresses responding and increases the threshold for responding (Van Wolfswinkel and Van Ree 1982).

Moreover, the endorphins seem to be implicated in the control of drug use. For example, blockade of opiate receptor systems with naltrexone decreased the reinforcing effects of ethanol in monkeys self-administering ethanol via the intravenous route (Altshuler *et al.* 1980). Thus beta-endorphin may be a common factor in drug use.

There has been recent speculation that endogenous opioids may mediate the reinforcing properties of cigarette smoking (Karras and Kane 1980; Tobin *et al.* 1982; Chernick 1983; Pomerleau *et al.* 1983). One supporting piece of evidence for the hypothesis is provided by a study which showed that plasma beta-endorphin levels were highly correlated with increases in plasma nicotine levels after smoking (Pomerleau *et al.* 1983). However, there has been no demonstration of any relation between brain endorphin and plasma endorphin.

The second prediction from the hypothesis is that endorphin blockers should reduce smoking. One study did find that pre-treatment with naloxone, the opioid antagonist, decreased cigarette smoking (Karras and Kane 1980). This led Nemeth-Coslett and Griffiths (1986) to test naloxone using an approach which had previously revealed dose-related sensitivity of cigarette smoking to the nicotine antagonist, meca-mylamine. If the hypothesis is correct, then naloxone would presumably block some of the reinforcing effects of cigarette smoking, like mecamylamine. However, naloxone administration did not significantly affect any of the behavioural or subject-rated measures of cigarette smoking. Specifically, naloxone pre-treatment did not significantly alter latency to first puff, number of cigarettes smoked, number of puffs taken, puff duration per cigarette, inter-puff interval, inter-cigarette interval, or time alight. Corresponding to the relative stability of cigarette smoking behaviour, carbon monoxide (CO) levels (post-session minus pre-session change scores, and post-session minus pre-drug change scores) did not differ across the different dose conditions.

The failure to replicate Karras and Kane's work is particularly notable because the study by Nemeth-Coslett and Griffiths used a similar subject population, examined a much wider range of naloxone doses which produced orderly dose-related subjective effects, used behavioural and biochemical measures of smoking, and used experimental procedures which had previously been shown to be sensitive for assessing drug effects on smoking (including the nicotine antagonist, mecamylamine). Thus, 'The failure of naloxone to produce any reliable effects on the multiple measures of smoking provides no support for the endogenous opioid theory of smoking reinforcement' (Nemeth-Coslett and Griffiths 1986).

Self-administration

A technique which has been shown to correlate with the euphoriant properties of a drug is self-administration, in which drug administration is contingent on the occurrence of a prior response. In a typical experiment, an animal is given access to a lever which delivers an intravenous injection of the drug. Using this technique, it has been shown that a variety of drugs from different pharmacological classes can serve as reinforcers in animal experiments (Schuster and Thompson 1969). Most of these drugs are abused by people, whereas those drugs that do not initiate and maintain responding in animals are in general not abused by people. Thus, the self-administration technique is a useful method to predict the abuse potential of drugs.

Endorphins support self-administration in rats when injected intraventricularly. In the study by Van Ree and his colleagues (1979), rats would work for an intraventricular injection when they received relatively low doses of beta-endorphin. Thus, it would be expected that it should be possible to set up an animal model of nicotine self-administration which closely resembles human self-administration. However, while there is a huge literature concerned with the self-administration of heroin, amphetamine, cocaine, and the barbiturates, the number of papers concerned with nicotine self-administration barely reaches double figures. This small number probably reflects the difficulty researchers experience in getting animals to self-inject the drug. Further, when animals do self-administer the drug, the amount varies considerably from day to day (Deneau and Inoki 1967). Monkeys can be trained to smoke although, only just over half the animals take up the habit, and once again the amount taken varies considerably from day to day (Jarvik 1967). Another feature which emphasizes the dissimilarity of this animal model to human smoking is the finding that when the researchers either stopped the delivery of tobacco smoke or switched the delivery to hot air, no immediate change was observed in the number of puffs the animals took from the tube which had delivered the smoke; although the puffing respose did extinguish over a few days in some animals, many did not alter their puffing rates over a fourteen-day period. Clearly, while nicotine must have some reinforcing properties, for animals this reinforcement is extremely weak in relation to other substances, and thus the self-administration does not closely resemble human smoking.

Finally, if nicotine were acting upon the same neural mechanisms as other habitually-used substances, we would expect that nicotine self-administration in people would be decreased by other habitually-used drugs. In people, use of amphetamine (Schuster *et al.* 1979), heroin (Mello *et al.* 1980), and alcohol increase cigarette smoking. Another study has shown that methadone produced dose-related increases in the cigarette smoking of methadone-maintained subjects (Chait and Griffiths 1984).

Subjective experience

The first supposition is that smoking would produce feelings of pleasure. However, in contrast to the intense feelings of pleasure which follow the administration of amphetamine or heroin, there are no sensations produced by nicotine or smoking which can be described as highly enjoyable in this way (Johnson 1942). Furthermore, not only are smokers who claim to smoke for pleasurable effects rather rare, but also those who do report that smoking increased positive affect only do so in terms of the enhancement of other rewarding experiences, for example the pleasurable relaxation of smoking after a meal.

In 1984, Henningfield published a widely-quoted study which was designed to assess 'abuse potential'. It involved the measurement of pleasure responses indicative of abuse potential following single doses of substances and gambling. The measure of pleasure was the Morphine Benzedrine Group scale of the Addiction Research Center Inventory. The subjects were individuals with a history of substance abuse. They were given a range of doses of the test compounds and placebo under double-blind conditions. Nicotine was given both intravenously and inhaled from tobacco smoke over a range of doses to eight subjects. Assessments were also made for other substances. Figure 3.1 shows the mean placebo scores in comparison with the scores at the highest dose tested of each substance on the Morphine Benzedrine Group scale of the Addiction Research Center Inventory.

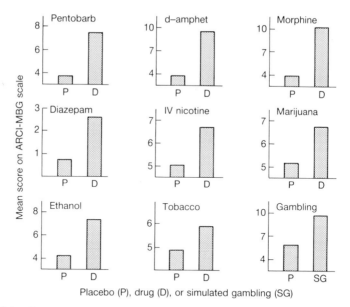

Placebo (P), drug (D), or simulated gambling (SG)

FIG. 3.1. Euphoriants for compulsive behaviours. A comparison of different substances and activities, using a Euphoriant scale from Henningfield (1984).

Henningfield (1984) concluded that 'The results of these studies provide direct evidence that nicotine, in doses comparable to those delivered by cigarette smoking, is an abusable drug. That is, nicotine meets the critical criteria of being psychoactive, producing euphoriant effects, and serving as a reinforcer. These findings suggest that the role of nicotine in cigarette smoking is similar to the roles played by other drugs in the maintenance of other kinds of substance self-administration, e.g. morphine in opium use, tetrahydrocannabinol (THC) in marijuana smoking, cocaine in coca leaf use, and ethanol in alcoholic beverage consumption'.

However, it should be noted that there are remarkable differences in the scales in Figure 3.1 and the results take on a different picture when plotted on the same scale (see Fig. 3.2). If the scores for the difference between assessment of the substance and placebo are plotted, as in Figure 3.3, we can derive a ranking of euphoria, or 'abuse potential' to use Henningfield's terms. It can be seen that nicotine injections and smoking are low on the euphoriant scale, and have low 'abuse potential' on the scale that Henningfield used for assessment.

On this evidence, there is not strong support for Henningfield's conclusion. From the data it seems that nicotine is, at best, a weak euphoriant and does not act like other compounds in the maintenance of

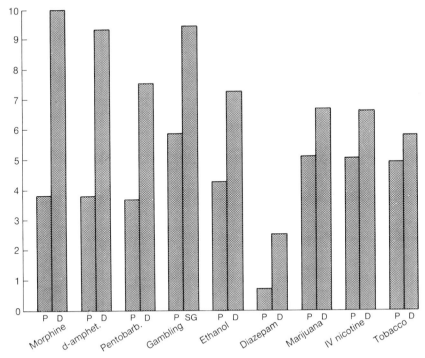

Placebo (P), drug (D) or simulated gambling (SG)

FIG. 3.2. Scores on the euphoriant scale for compulsive behaviour; the data from Figure 3.1 replotted on the same axis (data from Henningfield 1984).

other kinds of substance self-administration, i.e. it is not like morphine in opium use.

In the laboratory of the Department of Psychology, University of Reading, Henningfield's work has been followed up by considering the concepts of 'euphoria' and 'pleasurable well-being' in terms of two separable experiences, pleasurable-stimulation and pleasurable-relaxation. Tests were made on 139 subjects for their recall of their experience of different substances and activities. They were asked to rate these substances and activities on a scale from zero to ten, where ten was the maximum imaginable. The sample was a set of subjects who had some experience of using a variety of different substances. In this way, ratings of the pleasurable-stimulation and pleasurable-relaxation of a set of substances and activities, including tobacco, were obtained. These ratings enabled a comparative ranking of the substances and activities on these two kinds of euphoria to be derived (Fig. 3.4).

The most interesting comparisons were those of tobacco use with

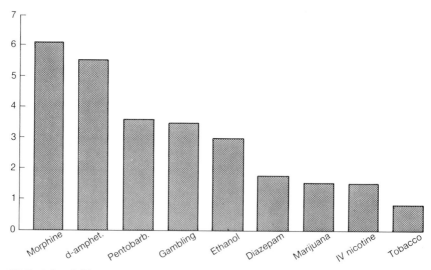

FIG. 3.3. Difference scores on the euphoriant scale for compulsive behaviour; the data from Figure 3.2 replotted as a differential from base line score (data from Henningfield 1984).

other substances and activities. Alcohol, amphetamines, amyl nitrite, cocaine, heroin, marijuana, and sex were significantly more stimulating than tobacco. Sleeping tablets and tranquillizers were significantly less stimulating than tobacco, while there were no statistically reliable differences between tobacco and caffeine or chocolate in terms of pleasurable-stimulation.

On the pleasurable-relaxation dimension, alcohol, heroin, sex, sleeping tablets, and tranquillizers were significantly more relaxing than tobacco. Amphetamine, amyl nitrite, cocaine, caffeine, and glue were significantly less relaxing than tobacco, while there were no statistically reliable differences between tobacco and chocolate in terms of pleasurable-relaxation.

The finding that smoking has pleasurable-stimulation and pleasurable-relaxing effects accords with the results of an early smoking-motives questionnaire by Ikard and his colleagues (1969). Positive-effect smoking had included the components stimulation, pleasurable relaxation, and manipulation. Some support for these factors has come from later work (Russell *et al.* 1974).

Summary

The findings on the Euphoriant Model of nicotine use strongly suggest

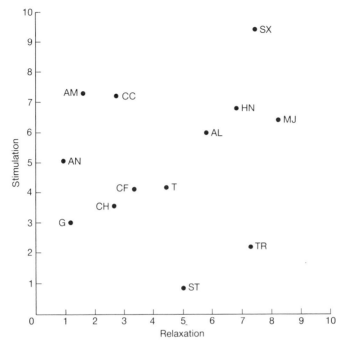

FIG. 3.4. A comparative ranking of substances and activities in terms of pleasure-stimulation and pleasure-relaxation.
Abbreviations: AL, alcohol; AM, amphetamines; AN, amyl nitrite; C, cocaine; CF, caffeine; CH, chocolate; G, glue; HN, heroin; MJ, marijuana; ST, sleeping tablets; SX, sex; T, tobacco; TR, tranquillizers.

that the pharmacological effects of nicotine are different from those of other habitually-used substances. Clearly, none of the predictions discussed are supported by experimental evidence. This must suggest that, whatever the nature of the mechanism by which nicotine maintains use, it is not the same as for the euphoriant mechanism activated by other habitually-used substances.

COPING MODEL

Since the beginning of time human beings have attempted to relieve psychological pain, dysphoria. Various coping strategies have been used for dysphoria, and one of the most common is drug administration. The vast bulk of modern psychopharmacology is devoted to examining the use of chemicals to alleviate psychological distress, with research work on anti-anxiety drugs, antidepressant drugs, and antipsychotic drugs. In this section, the use of nicotine as an agent for coping with psychological distress is considered.

The Coping Model has been implicit in the writings of a number of researchers (Eysenck 1965; Stepney 1980; Warburton and Wesnes 1978). This view sees smoking as a response to a pre-existing distress or deficiency of some sort. The pre-existing factors could be constitutional, i.e. a neural inadequacy, a personality trait (biologically- or socially-determined), and situational or social problems. Whatever the source, smoking is seen as a coping strategy for the problem. This is a formalization of ideas that have been discussed previously in a series of papers from the laboratory of the University of Reading (Warburton and Wesnes 1978, 1983; Warburton *et al.* 1983; Warburton 1985) in which smoking was described as a form of 'self-medication', a coping strategy for every-day problems. This view is similar to that of Alexander and Hadaway (1982). In their paper on opiate use, they discuss adaptive orientation, which is the idea that habitual opiate use is an attempt to adapt to chronic distress of any sort through use of opiate drugs.

Personality determinants of smoking

The largest personality surveys of smokers were made by Eysenck (1963) and his colleagues (1960). These studies of male smokers gave no evidence for a significant correlation of smoking with neuroticism (Eysenck 1963), and Rae's 1975 survey of 253 female students revealed no differences in the degree of neuroticism between non-smokers, ex-smokers, 'light' smokers, or medium smokers (less than 15 per day). Nevertheless, six studies have suggested a positive relationship of smoking with neuroticism, and this association seems particularly strong for women smokers (Guilford 1966; Meares *et al.* 1971; Waters 1971; Dunnell and Cartwright 1972; Shiffman 1979*b*; Warburton *et al.* 1983).

It has been argued that repeated exposure to nicotine caused anxiety, and thus a higher neuroticism score (Schachter 1978). Evidence against this argument comes from a prospective study done by Cherry and Kiernan (1978). At the age of 16 years, 2853 young people completed the Maudsley Personality Inventory before most of them had begun to smoke. At 20 years and 25 years of age they completed a smoking-habits questionnaire, and it was found that the cigarette smokers, as a group, scored more highly on neuroticism. This finding argues strongly that constitutional factors underlying neuroticism determine whether some people smoke.

Situational determinants

In a study of smoking situations, McKennell (1970) classified smokers according to the occasions on which they thought that they were likely to smoke. The answers to the questions were factor-analysed, and seven smoking situations were found;

(1) 'nervous irritation smoking';

(2) 'relaxation smoking';

(3) 'smoking alone';

(4) 'activity-accompanying smoking';

(5) 'food substitution smoking';

(6) 'social smoking';

(7) 'social confidence smoking'.

The nervous irritation factor included 'smokes when anxious or worried', and 'smokes when nervous', and also 'smokes when angry'. When cigarette consumption was equated, female smokers had higher scores for nervous irritation smoking. The first five types of smoking were correlated and formed an 'inner need factor', an internal determinant. The last two factors were associated, and could be identified as 'social factors', situational-determinant.

Another study of situational smoking was done by Frith (1971). He asked smokers to rate their desire for a cigarette in a set of imagined situations. A principal-components analysis of their answers had a positive loading on desire to smoke in each situation, as well as on cigarette consumption, while the second component contrasted desire to smoke in high and low arousal-inducing situations, i.e. sedative and stimulant smoking. Frith subdivided his proposed high arousal-inducing situations into those involving emotional stressors and anxiety, and those in which the stressors could be due to mental activity. Women were more likely than men to smoke in stressful, high-arousal situations. The low-arousal situations focused on relaxation, boredom, repetitive work, and fatigue.

Russell and his colleagues (1974) combined some McKennell items and some of the Ikard *et al.* (1969) items into a single questionnaire. The outcome of the factor analysis were six factors which they labelled as psychosocial, indulgent, sensorimotor, stimulation, addictive, and automatic. On the item analysis, they found that 93 per cent of smokers who attended a smoking clinic and 74 per cent of non-clinic smokers said that they smoked when worried, while Warburton and Wesnes (1978) found that 88 per cent of students answered in the same way. It was interesting that there are differences in female and male motives for smoking; it is said that women smoke more to relieve anxiety and anger than men (Russell *et al.* 1974). These data fitted very neatly with the findings of McKennell and Frith on smoking motives and smoking situations.

Combination of the questionnaires was carried a stage further by Stanaway and Watson (1980) who used all the items of Russell's smoking questionnaire with all the items of Frith's situational smoking questionnaire, and data on cigarette consumption, the nicotine yield of the preferred cigarette brand, reported depth of inhalation, and the age at which they started to smoke. In a two-factor solution, one factor grouped the 'high-stress' factor from Frith's situational smoking questionnaire with the pharmacological dependence factor in Russell's smoking questionnaire. The other factor combined Frith's 'low-stress' factor with items from the non-pharmacological factor from Russell's questionnaire, the pleasurable relaxation items, and some of the sensory items. It appeared that the pharmacological motive for smoking was much more closely related to smoking in stressful situations than in relaxing situations.

Smoking cessation

Evidence for the importance of smoking in coping with anxiety comes from smoking cessation studies. Around 25 per cent of smokers stop without any symptoms. However, one of the commonest reported symptoms is anxiety, which is more likely to be reported by women (Guilford 1966; Shiffman 1979a). Analysis of the situations which resulted in a return to smoking (by Marlatt 1979, and Shiffman 1979b) indicated that 80 per cent of these situations fell into three categories; coping with anxiety and other negative emotional states (43%), social pressure (25%), and coping with social stress (12%). In the Shiffman 1982 study, two-thirds of the subjects were under stress at the time of relapse, and anxiety was particularly common among ex-smokers who relapsed at work, which suggests that work-related anxiety was a contributory factor. Smokers who have high degrees of neuroticism are more anxiety-prone, and so it is not surprising that they find it difficult to stop smoking, or relapse if they abstain (Cherry and Kiernan 1978). These data on smoking to cope are congruent with studies showing that deprived smokers are more likely to feel anxious in stressful situations (Schachter 1978).

Smoking and stressful situations

In order to determine whether smokers smoke more in times of stress, the smoking habits of students during examinations has been studied (Warburton *et al.* 1983). Students kept a diary of their smoking during an examination week and then during the less stressful summer term. The students smoked more during the examination period than during the non-examination period. They smoked more cigarettes on mornings preceding an afternoon examination, over 80 per cent more than during a non-examination period. In fact only two of the 48 subjects did not increase their morning cigarette consumption during the examination period. Throughout the examination period the students reported that they inhaled more strongly, and to a greater depth, than during the non-examination period.

The cigarette butts were analysed for nicotine content; the estimated yield of nicotine to the mouth was slightly lower during the examination period. However, the product of the mouth nicotine levels and the number of cigarettes was greater during the examination period. Even if the subjects had not stated that they had inhaled more of the smoke they generated, it means that they must have increased their nicotine intake during this stressful time.

Experimental studies of mood

Experimental studies of mood are difficult to conduct, and not many satisfactory ones exist. Most of them have used noise as a stressor. In one early example, Schachter (1978) reports a study by Perlick which compared irritability in 'unrestrained' smokers and 'restrained' smokers who were trying to cut down their smoking. Each group rated the annoyance caused by aircraft noise, after smoking a 1.3 mg nicotine cigarette, smoking a 0.3 mg nicotine cigarette, or not smoking. Unrestrained smokers, when deprived or smoking the 0.3 mg nicotine cigarette, were more annoyed than when smoking the 1.3 mg cigarette. However, unrestrained smokers, when smoking the 1.3 mg cigarette, were neither less nor more irritated than non-smokers. The restrained group were just as irritated in all three conditions as the deprived or low nicotine unrestrained smokers. These data are consistent with the hypothesis that smokers are constitutionally more irritable but nicotine from smoking helps to reduce this feeling, and allows a subject to cope more successfully with the stressor.

In a series of studies, Mangan and Golding (1978) tested the effects of smoking on autonomic responsivity as measured by skin conductance to brief neutral noise stimuli and to aversive bursts of white noise. In comparison with sham smoking, smoking reduced the response to the stress stimuli; it modulated the arousal response, which thus became less 'stress-like'.

More evidence which is consistent with the reduction of stress hypothesis comes from a study that compared the effects of smoking with sham smoking during intermittent noise bursts (Woodson *et al.* 1986). Subjective stress responses increased in the sham smoking condition but not during real smoking. In addition, smoking decreased the physiological stress response as measured by electrodermal activity, heart rate changes, and vasoconstriction. It seems that smoking dampens down and smooths out the body's stress response to these noises. This latter finding agreed with that of an earlier study by Gilbert and Hagen (1980), which found that smoking reduced electrodermal activity during the viewing of emotional films depicting mutilation, snakes, fights, anger, and fear.

Aggression in smokers was studied by Cherek (1981) in a 'test' in which subjects were told that they could accumulate money by their efficiency of responding in an information-processing task which involved response rate, reaction time, etc. Subjects were misinformed that they had been paired with another person who was able to subtract money from them, and who might annoy them with blasts of white noise. They were also told that they could reduce the amount of money earned by the other person and give them blasts of white noise, if they wished. Counts of the number of these behaviours constituted measures of aggression.

Provocation of aggressive responses in the subject was produced by the loss of money and blasts of white noise from their fictitious partner. Smoking helped the smokers to be less aggressive when they believed that they were being cheated by the other person in the test. Higher yield cigarettes (2.19 mg) were more effective in reducing aggressive responding than lower yield products (0.42 mg). The decreased aggressive behaviour could have been due to the suppression of the effects of nicotine withdrawal or to the fact that smokers are constitutionally more aggressive and use smoking to cope with aggression. The constitutional explanation is more likely, given the reduction of aggression in animals by nicotine (Driscoll and Battig 1981; Rodgers 1979; Silverman 1971).

In a study intended to study anxiety, Silverstein (cited by Schachter 1978) assessed the amount of shock which could be tolerated by smokers who were deprived, or smoking either a 0.3 mg nicotine cigarette or a 1.3 mg cigarette. Silverstein argued that the more anxious the subjects the less pain they would be able to tolerate. Smoking enabled smokers to stand higher levels of electric shock than they could when not smoking. However, deprived smokers were poorer than non-smokers in pain tolerance, and smoking the higher-yield product only enabled smokers to stand the same levels as non-smokers. This finding is consistent with the Coping Model, in which smokers are constitutionally less tolerant; smoking only helps them function like people who do not smoke. It would also be consistent with the hypothesis that smokers who are deprived suffer withdrawal symptoms and they need cigarettes to be able to function.

In a study of the puff-by-puff mood effects of cigarettes at the University of Reading, subjects completed a set of Bond-Lader mood scales after each puff. Subjects reported that they became calmer, more relaxed, more tranquil, more sociable, more friendly, more contented, and happier as they smoked the cigarettes. The changes on the 'relaxation' set of scales were curvilinear, while the puff-by-puff changes on the 'contentedness' scale were linear.

Plasma nicotine was studied while smoking middle-tar and low-tar cigarettes, similar to those in the mood study. When measured puff-by-puff, increases in plasma nicotine were variable for most subjects; different asymptotes were reached at the end of the cigarettes, but no significant difference was found, due to variation. The plasma nicotine changes were correlated with mood; as plasma nicotine increased throughout the cigarette, the mood changes increased.

Smoking for stimulation

It is also consistent with the Coping Model that smokers should differ in their personality with respect to stimulation. Personality surveys of smokers by Eysenck (1963) and his colleagues (1960) have revealed a high positive correlation between amount of cigarette consumption and degree of extraversion. Studies on both sexes by Warburton and Wesnes (1978) have confirmed this statistical relationship between smoking and extraversion.

As mentioned earlier, Cherry and Kiernan (1978) found that smokers scored more highly on neuroticism, but they also had a higher degree of extraversion. The two personality dimensions were independent and additive as predictors of becoming an habitual smoker. This finding argues strongly that constitutional factors underlying extraversion, as well as neuroticism, determine whether some people smoke.

The dominant characteristic of people scoring high on the extraversion scale is that they are under-aroused and so have a need for stimulation, which explains their lifestyle. They get bored easily and have poorer concentration. Smokers overwhelmingly say that 'Smoking helps me to relax', and 'Smoking helps me to think and concentrate' (Russell *et al.* 1974; Warburton and Wesnes 1978). In the studies at the University of Reading, 74 per cent of smokers report that this help is a motivation to smoke (Wesnes *et al.* 1984).

Given the above data, the same team investigated whether smoking has beneficial effects on concentration in smokers.

Experimental studies of stimulation

First studies found that smoking helped smokers to maintain their performance in a prolonged vigilance task, whereas performance decreased markedly for non-smokers, and also for a group of smokers who were not allowed to smoke during the test. This maintenance of performance occurred for both visual and auditory vigilance tasks. In the auditory vigilance test, the decrement was relative to smoking nicotine-free cigarettes (Wesnes and Warburton 1978).

In another series of studies, Wesnes and Warburton examined rapid, visual information-processing performance both before and after smoking, in a task in which the speed and the accuracy of processing were measured. In the first test, a high-nicotine cigarette produced a greater increase in both speed and accuracy than lower delivery cigarettes (Wesnes and Warburton, 1983*a*). In a second study, smoking improved speed and accuracy above baseline levels whereas either not smoking or smoking nicotine-free cigarettes resulted in a decline in speed and accuracy below baseline levels (Wesnes and Warburton 1983*b*).

In a third study in this series, not only did smoking improve the speed and accuracy above both pre-smoking levels and above that following not smoking, but it was also found that higher-delivery cigarettes improved performance more than low-delivery ones (Wesnes and Warburton 1984*b*). This task was used to study the performance effects while the smokers were smoking a cigarette, it was found that the performance was improved during smoking, and that these improvements persisted throughout the cigarette. These improvements during smoking were greater than those found after smoking. After smoking, the improvements were about seven to eight per cent, while during smoking they were 14 to 15 per cent.

These studies provide evidence that smokers perform more efficiently after smoking than after not smoking, that this improvement is made in terms of both speed and accuracy, and that higher-nicotine products improve performance more than low-nicotine or nicotine-free products.

Summary

There is strong support for the Coping Model for smoking behaviour. People believe that smoking does improve their mood, and there is experimental evidence to support this idea. Smokers increase their nicotine intake at stressful times.

FUNCTIONAL MODEL

The functional view of smoking has been described by Warburton

(1985). The functional approach regards smoking as a person's use of nicotine to control their psychological state. The model is purposive; it views smoking as a strategy for managing everyday needs. The reasons for smoking pre-date its initiation, and smoking is maintained because it satisfies these needs, which are due to both exogenous and endogenous causes, situational as well as constitutional. Consequently, smokers smoke for different reasons. Smoking is a multi-dimensional behaviour which must be conceptualized by a multifactorial model to explain all the reasons for smoking.

This view is a formalization of ideas that have been discussed previously in a series of papers from the laboratory of the Department of Psychology, University of Reading (Warburton 1979, 1985; Warburton and Wesnes 1978, 1983; Warburton *et al.* 1983) and other laboratories (Russell 1971; Ashton and Stepney 1982). It encompasses the idea that smoking is used as a form of 'self-medication', a coping strategy for every-day problems. However, the Functional Model is a more general form of this view without the quasi-medical connotations. The important aspect of the model is that there need not be any psychological distress to obtain the benefits of smoking, but that the non-smoker could equally benefit from nicotine if smoking were acceptable in aesthetic and health concerns. The evidence for the broader Functional Model, as opposed to the simpler Coping Model, comes from studies of nicotine with light smokers, but especially from non-smokers. The Coping Model would not predict that 'heavy' smokers, 'light' smokers, and non-smokers would benefit to the same extent from the amount of nicotine that is usually obtained from smoking a cigarette.

Nicotine and performance

In the visual vigilance task (described earlier), smoking reduced the vigilance decrement in the non-smoking condition for smokers (Wesnes and Warburton 1978). A follow-up study (Wesnes *et al.* 1983) investigated the effects of nicotine tablets on the vigilance performance of non-smokers, 'light' smokers (who averaged less than five cigarettes per day), and 'heavy' smokers (who used more than 15 per day). Some of the 'light' smokers only smoked at weekends; none of them usually smoked in the mornings, and none of them complained about abstaining from cigarettes during the 12 hours preceding the experiment, which was not the case for the 'heavy' smokers.

Comparisons of the vigilance performance indicated that the differences between the placebo and nicotine tablets were significant for both groups, and that there was no difference in the effect of nicotine on the performance of 'light' and 'heavy' smokers. If nicotine withdrawal were the primary factor in determining the effects of nicotine on performance, then one would expect 'heavy' smokers to be affected more by nicotine than the 'light' smokers. If tolerance to nicotine were the primary factor in determining the effects of nicotine on performance, then one would expect 'heavy' smokers to be less affected by nicotine than the 'light' smokers. There was no evidence for either view.

In another study (Wesnes and Warburton 1984*a*), nicotine tablets were given to non-smokers, who then performed the rapid, visual information-processing task used to study cigarette smoking in the previous section. We found that after the 1.5 mg tablet, the performance improvement was similar to that produced by smoking. Thus, this study provides strong evidence that nicotine plays a major role in the improvements found with cigarette smoking on this task.

The effects of nicotine tablets were also tested using the Stroop Test. In the Stroop Test, a subject names the print colour of a series of colour names, but the colour names never correspond to the colour of the ink in which they are printed. This incongruity is very distracting, and subjects take longer to name this list than an equivalent length list of colour patches. The time difference between naming the colours in the two conditions is the Stroop effect. It is a measure of the distraction caused by the incongruent colour words, a failure of attention.

The effects of 0, 1, and 2 mg of nicotine were studied on the Stroop performance of smokers and non-smokers. Nicotine reduced the magnitude of the Stroop effect in this study. No differences were found either between the effects of the two doses of nicotine on the Stroop scores or between the responsivity of the smokers and the non-smokers to the drug. The latter finding adds support to the argument that the smokers and non-smokers do not differ in their response to nicotine. Consequently, it cannot be argued that the 'heavy' smokers need nicotine in order to perform 'normally' or that the 'light' smokers were more sensitive to nicotine.

CONCLUSION

As has been pointed out elsewhere (Warburton and Wesnes 1983; Warburton 1987) there are many motives for smoking.

The Euphoriant Model argues that nicotine use is maintained by its effect on the pleasure systems in the brain. However, nicotine does not act on the reward pathways in the brain or on the neurochemical systems that have been associated with euphoria. It has proved very difficult to get animals to self-administer nicotine, and smokers have shown that they do not actually experience any marked euphoriant effects. Thus, whatever the nature of the mechanism by which nicotine maintains use, it is not the same as that activated by other substances which are habitually used.

The Coping Model of nicotine use proposes that nicotine is used to alleviate negative psychological states. Studies of the personality of smokers, the situational determinants of smoking, smoking cessation, smoking behaviour under stress, and experiments on mood provide strong support for the Coping Model for smoking behaviour. However, the Functional Model does not require a single motive for all smokers. People may smoke for different reasons on different occasions. The level of smoking will reflect the personal function of smoking for the individual. The beneficial effects result from the functions that smoking serves for the individual. Smoking can be seen as being maintained by the personal control that smokers have over their psychological state because of a very effective delivery of nicotine to the brain, which enables virtually instant adjustment of the person's psychological state.

The Functional Model sees smoking as an important resource for the person, i.e. something that is available to them for managing their lives. People have a range of resources that are available for the management of their daily lives, of which smoking is but one, and the management of everyday life cannot be explained in terms of a single resource. The word 'managing' has been used rather than 'coping', since managing incorporates fewer normative connotations than coping. Life is not just a series of pathogenic events and problems for the individual to cope with or not, as the case may be. Coping implies drama, and 'failure to cope' becomes labelled maladaptive. The Functional Model sees smoking as a resource for managing everyday life, not just for coping with problems.

This functional view of smoking sees it as the outcome of the interaction of the person and the situation. The personality characteristics of the smoker are critical in this interaction, and thus in the level of smoking. In the final analysis, smoking delivers nicotine to the brain, and smoking can be used by the smoker to obtain nicotine which provides extra resources for the smoker. Thus, smoking is a purposeful activity for smokers; it provides them with a resource for managing their lives.

REFERENCES

Alexander, B. K. and Hadaway, P. F. (1982). Opiate addiction: The case for an adaptive orientation. *Psychol. Bull.* **92**, 367–81.

Altshuler, H. L., Phillips, P. E., and Feinhandler, D. A. (1980). Alteration of ethanol self-administration by naltrexone. *Life Science* **26**, 679–88.

Ashton, H. and Stepney, R. (1982). *Smoking: psychology and pharmacology.* University Press, Cambridge.

Blum, K. (1984). *Handbook of abusable drugs.* Gardner Press, New York and London.

Chait, L. D. and Griffiths, R. R. (1984). Effects of methadone on human cigarette smoking and subjective ratings. *J. Pharmac. and Exp. Ther.* **229**, 636–40.

Cherek, D. R. (1981). Effects of smoking different doses of nicotine on human aggressive behavior. *Psychopharmacology* **75**, 339–45.

Chernick, V. (1983). The brain's own morphine and cigarette smoking: the junkie in disguise? *Chest* **83**, 2–4.

Cherry, N. and Kiernan, K. (1978). A longitudinal study of smoking and personality. In *Smoking behaviour: physiological and psychological influences* (ed. R. E. Thornton) pp. 12–18. Churchill-Livingstone, Edinburgh.

Crow, T. J. (1970). Enhancement by cocaine of intra-cranial self-stimulation in the rat. *Life Science* **9**, 375–81.

Deneau, G. A. and Inoki, R. (1967). Nicotine self-administration in monkeys. *Annals of the New York Academy of Science* **142**, 277–9.

Domino, E. F. (1973). Neuropsychopharmacology of nicotine and tobacco smoking. In *Smoking behavior: motives and incentives* (ed. W. L. Dunn). Winston, Washington DC.

Driscoll, P. and Battig, K. (1981). Selective inhibition by nicotine of shock-induced fighting in the rat. *Pharmacology, Biochemistry and Behavior* **14**, 175–9.

Dunnell, K. and Cartwright, A. (1972). *Medicine takers, prescribers and hoarders.* Routledge and Kegan Paul, London.

Eysenck, H. J. (1963). Personality and cigarette smoking. *Life Science* 3 **29**, 777–92.

—— (1965). *Smoking, health and personality.* Weidenfeld and Nicolson, London.

——, Tarrant, M., Woolf, M., and England, L. (1960). Smoking and personality. *Brit. Med. J.* **1**, 1456–60.

Frith, C. D. (1971). Smoking behaviour and its relationship to the smoker's immediate experience. *Brit. J. Soc. Clin. Psychol.* **10**, 73–8.

Fuxe, K., Agnati, L., Eneroth, P., Gustafsson, J-A., Hokfelt, J., Lofstrom, A., Skett, B., and Skett, P. (1977). The effect of nicotine on central catecholamine neurones and gonadotrophin secretion. *I. Med. Biol.* **55**, 148–76.

Gilbert, D. G. and Hagen, R. L. (1980). The effects of nicotine and extraversion on self-report, skin conductance, electromyographic, and heart responses to emotional stimuli. *Addictive Behaviors* **5**, 247–57.

Guilford, J. S. (1966). Factors related to successful abstinence from smoking. Pittsburgh, American Institutes for Research.

Henningfield, J. E. (1984). Pharmacological basis and treatment of cigarette smoking. *J. Clinical Psychiatry* **45**, 24–34.

Ikard, F. F., Green, D. E., and Horn, D. A. (1969). A scale to differentiate between types of smoking as related to management of affect. *Int. J. Addictions* **4**, 649–59.

Jarvik, M. E. (1967). Tobacco smoking in monkeys. *Ann. NY Acad. Sci.* **142**, 280–94.

—— (1973). Further observations on nicotine as the reinforcing agent in smoking. In *Smoking behavior: motives and incentives* (ed. W. L. Dunn). Wiley, New York.

Johnson, L. M. (1942). Tobacco smoking and nicotine. *Lancet* **2**, 742.

Karras, A. and Kane, J. M. (1980). Naloxone reduces cigarette smoking. *Life Science* **27**, 1541–5.

Koob, G. F., Spector, N. H., and Meyerhoff, J. L. (1975). Effects of heroin on lever-pressing for intracranial self-stimulation, food and water in the rat. *Psychopharmacologia* **42**, 231–4.

Lorens, S. A. and Mitchell, C. L. (1973). Influence of morphine on lateral hypothalamic self-stimulation in the rat. *Psychopharmacologia* **32**, 271–7.

McKennell, A. C. (1970). Smoking motivation factors. *Brit. J. Soc. Clin. Psychol.* **9**, 8–22.

Mangan, G. and Golding, J. (1978). An 'enhancement' model of smoking maintenance? In *Smoking behaviour: physiological and psychological influences* (ed. R. E. Thornton) pp. 115–26. Churchill and Livingstone, Edinburgh.

Marlatt, A. (1979). A cognitive-behavioral model of the relapse process. In *Behavioural analysis and treatment of substance abuse* (ed. N. A. Krasnegor) pp. 191–9. National Institute for Drug Abuse, Washington DC.

Meares, R., Grimwade, J., Bickley, M., and Wood, C. (1971). Smoking and neuroticism. *Lancet* **2**, 770.

Mello, N. K., Mendelson, J. H., Sellers, M. L., and Kuehnle, J. C. (1980). Effects of heoroin self-administration on cigarette smoking. *Psychopharmacology* **67**, 45–52.

Mogenson, G. J. (1964). Effects of sodium pentobarbital on brain self-stimulation. *J. Comp. Physiol. Psychol.* **58**, 461–2.

Nemeth-Coslett, R. and Griffiths, R. R. (1986). Naloxone does not affect cigarette smoking. *Psychopharmacology* **89**, 261–4.

Pomerleau, O. F., Fertig, J. B., Seyler, L. E., and Jaffe, J. (1983). Neuroendocrine reactivity to nicotine in smokers. *Psychopharmacology* **81**, 61–7.

Rae, G. (1975). Extraversion, neuroticism and cigarette smoking. *Brit. J. Soc. Clin. Psychol.* **14**, 429–30.

Rodgers, R. J. (1979). Effects of nicotine, mecamylamine and hexamethonium on shock-induced fighting, pain reactivity and locomotor behavior in rats. *Psychopharmacology Berlin* **66**, 93–8.

Russell, M. A. H. (1971). Cigarette smoking: natural history of a dependence disorder. *Brit. J. Med. Psychol.* **44**, 1–16.

——, Peto, J., and Patel, U. A. (1974). The classification of smoking by factorial structure of motives. *J. Roy. Stat. Soc. A.* **137**, 313–33.

St Laurent, J. and Olds, J. (1967). Alcohol and brain centers of positive reinforcement. In *Alcoholism: behavioral research, therapeutic approaches* (ed. R. Fox) pp. 85–106. Springer-Verlag, New York.

Schachter, S. (1978). Pharmacological and psychological determinants of smoking. In *Smoking behaviour: physiological and psychological influences* (ed. R. E. Thornton) pp. 208–28. Churchill-Livingstone, Edinburgh.

Schuster, C. R. and Thompson, T. (1969). Self-administration of and behavioral dependence on drugs. *Ann. Rev. Pharmacol.* **9**, 483–502.

——, Lucchesi, B. R., and Emley, M. S. (1979). The effects of d-amphetamine, meprobamate and lobeline on the cigarette smoking behavior of normal human subjects. In *Cigarette smoking as a dependence process* (ed. N. A. Krasnegor) pp. 91–9. National Institute for Drug Abuse, Washington DC.

Shiffman, S. M. (1979*a*). The tobacco withdrawal syndrome. In *Cigarette smoking* *as a dependence process* (ed. N. A. Krasnegor). National Institute for Drug Abuse, Washington DC.

—— (1979*b*). Analysis of relapse episodes following smoking cessation. Paper presented at the 4th World Congress on Smoking and Health.

—— (1982). Relapse following smoking cessation. *J. Consult. Clin. Psychol.* **50**, 71–86.

Silverman, A. P. (1971). Behavior of rats given a smoking dose of nicotine. *Anim. Behav.* **19**, 67–72.

Stanaway, R. G. and Watson, D. W. (1980). Smoking motivation: A factor-analytical study. *Personal. Ind. Diff.* **1**, 371–80.

Stein, L. (1962). Effects and interactions of imipramine, chlorpromazine, reserpine and amphetamine on self-stimulation: Possible neurophysiological basis of depression. In *Recent advances in biological psychiatry* (ed. J. Wortis) pp. 288–308. Plenum, New York.

Stepney, R. (1980). Smoking behavior: A psychology of the cigarette habit. *Brit. J. of Diseases of the Chest* **74**, 325–44.

Tobin, M. J., Jenouri, G., and Sachner, M. A. (1982). Effects of naloxone on change in breathing pattern with smoking: A hypothesis on the addictive nature of cigarette smoking. *Chest* **82**, 530–7.

Van Ree, J. M., Smyth, D. G., and Colpaert, F. (1979). Dependence creating properties of lipotropin C-fragment. (Beta-endorphin): Evidence for its internal control of behavior. *Life Science* **24**, 495–502.

Van Wolfswinkel, L. and Van Ree, J. M. (1982). Effects of morphine and naloxone on ventral tegmental electrical self-stimulation. In *Drug discrimination: Application in CNS pharmacology* (eds. F. C. Colpaert and J. L. Slangen) pp. 391–7. Elsevier, Amsterdam.

Warburton, D. M. (1975). *Brain, behaviour and drugs.* Wiley, London.

—— (1979). Self medication. In *Psychology and medicine* (eds. D. J. Oborne, M. M. Gruneberg, and J. R. Eiser) pp. 196–204. Academic Press, London.

—— (1985). Nicotine and the smoker. *Rev. Environ. Health* **5**, 343–90.

—— (1987). The functions of smoking. In *Tobacco smoke and nicotine: a neurobiological approach*, pp. 178–99. Plenum Press, New York.

—— and Wesnes. K. (1978). Individual differences in smoking and attentional performance. In *Smoking behaviour: physiological and psychological influences* (ed. R. E. Thornton) pp. 19–43. Churchill-Livingstone, Edinburgh.

—— and —— (1983). Mechanisms of habitual substance use: food, alcohol and cigarettes. In *Physiological correlates of human behaviour* (eds. A. Gale and J. Edwards) pp. 277–98. Academic Press, London.

——, ——, and Revell, A. (1983). Personality factors in self-medication by smoking. In *Response variability to psychotropic drugs* (ed. W. Janko, in press). Pergamon Press, London.

Waters, W. E. (1971). Smoking and neuroticism. *Brit. J. Prevent. Soc. Med.* **25**, 162–4.

Wesnes, K. and Warburton, D. M. (1978). The effects of cigarette smoking and nicotine tablets upon human attention. In *Smoking behaviour: physiological and psychological influences* (ed. R. E. Thornton) pp. 131–47. Churchill-Livingstone, Edinburgh.

—— and —— (1983*a*). Smoking, nicotine and human performance. *Pharmacol. Therap.* **21**, 189–208.

—— and —— (1983*b*). The effects of smoking on rapid information processing performance. *Neuropsychobiol* **95**, 223–9.

—— and —— (1984*a*). Effects of scopolamine and nicotine on human rapid information-processing performance. *Psychopharmacol* **82**, 147–50.

—— and —— (1984*b*). The effects of cigarettes of varying yield on rapid information processing performance. *Psychopharmacology* **82**, 338–42.

——, ——, and Matz, B. (1983). The effects of nicotine on stimulus sensitivity and response bias in a visual vigilance task. *Neuropsychobiology* **9**, 41–4.

——, ——, and Revell, A. (1984). Work and stress are motives for smoking. In *Smoking and the lung* (eds. G. Cumming and G. Bonsignore) pp. 233–48. Plenum Press, New York.

Wise, R. A. (1982). Neuroleptics and operant behavior: The anhedonia hypothesis. *Behav. Brain Sci.* **5**, 39–53.

Woodson, P. P., Buzzi, R., Nil, R., and Battig, K. (1986). Effects of smoking on vegetative reactivity to noise in women. *Psychophysiology* **23**, 272–82.

4

Ethanol, phosphoinositides, and transmembrane signalling—towards a unifying mechanism of action

ERIK ÄNGGÅRD

INTRODUCTION

Ethanol (ethyl alcohol) is the oldest and most widely used psychotropic agent. Its primary effects are mood elevation, anxiety relief and, at higher doses, sedation. People drink ethanol largely to experience these effects, which is the reason why ethanol is addictive. The secondary effects of alcohol are those which are related to changes in liver metabolism. Even at low concentrations of ethanol, a major part of the liver function is devoted to the metabolism of ethanol.

Clearly it is the primary effects of ethanol which cause alcoholism, whereas the secondary effects of ethanol are associated with alcohol-related diseases, such as liver cirrhosis and cardiomyopathy. In this chapter the primary effects of ethanol at the molecular and cellular level will be discussed, and a unifying mechanism of action proposed.

It is only very recently that there has been some insight into the fundamental mechanism of action of ethanol on the brain. Two observations appear to be of particular significance.

1. Ethanol increases the fluidity of cell membranes.

2. Ethanol inhibits stimulus secretion/contraction coupling.

These findings can now be rationalized in the light of new studies showing effects of ethanol on membrane phosphoinositide metabolism.

ETHANOL AND MEMBRANE FLUIDITY

There is general agreement that the fundamental action of ethanol, as well as other anaesthetics, is to change the properties of the neuronal cell membrane (for review, see Taraschi and Rubin 1985).

According to present views, the plasma membrane consists of a bimolecular layer in which the hydrophilic groups of phospholipid molecules are directed to the surface, while vertically arranged fatty acids are directed to the interior of the membrane (Fig. 4.1, top). Proteins are inserted into this bilayer and penetrate partially or completely. The fatty acids have a broad range of softness, which defines the fluidity of the membrane. Saturated fatty acids are very stiff at body temperature, whereas unsaturated fatty acids with many double bonds are very mobile and confer greater fluidity.

The synapse requires high fluidity. In neurotransmission, synaptic vesicles are fused with the presynaptic membrane and the neurotransmitter released in the synaptic cleft. Analysis of synaptic membranes has shown them to be rich in soft, polyunsaturated fatty acids, and to have a lower proportion of stiff lipids.

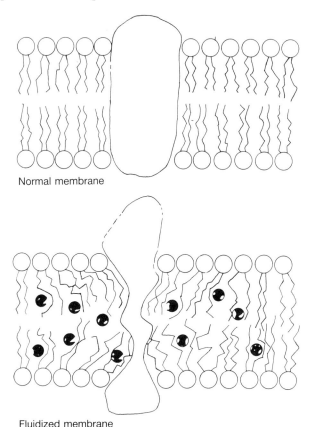

Normal membrane

Fluidized membrane

FIG. 4.1. Schematic illustration showing normal cell membrane (top) and cell membrane fluidized by the presence of ethanol molecules. Note distortion of integral movements in the cell membrane fatty acyl groups.

Ethanol produces its effect on the brain by entering neuronal membranes and changing their fluidity (Fig. 4.1, bottom). Thus it behaves very much like other central depressants and anaesthetics. For all such drugs it has been found that about 500 millimoles per 100 gram of membranes is necessary to induce anaesthesia. Increasing disorder, e.g. fluidization, has been observed in brain synaptosomal membranes in ethanol concentrations between 20 and 350 mM (Chin and Goldstein 1977).

The evidence for the fluidizing effect of ethanol on cell membranes has been obtained using different techniques, e.g. studying electrophysiology of nerve membranes, properties of proteins in cell membranes, and physical measurements (Taraschi and Rubin 1985). The strongest evidence comes from the use of spin-labelled fatty acids incorporated into the membranes as probes, and the use of electron spin resonance to measure changes in fluidity. From these experiments it has been possible to conclude that ethanol acts on the middle of the bimolecular membrane, and also that the ethanol effect is greater on membranes which are already fluid (Fig. 4.1, bottom).

It is important to recognize that there may be particularly vulnerable microdomains in the membrane with high basal fluidity, e.g. in synapses and around the membrane proteins. Thus the shape, and therefore the activity, of a membrane-bound enzyme may be influenced by changes in the fluidity of the lipids surrounding it.

ETHANOL INHIBITS STIMULUS SECRETION (SSC) CONTRACTION

The second fundamental mechanism of action is inhibition of neuro-transmitter release and stimulus secretion/contraction coupling, SSC (Kalant and Grose 1967; Carmichael and Israel 1975). This is the collective name for a chain of events beginning with the activation of the receptor on the outside of an effector cell leading to an effect either in the form of extrusion of cell content in the form of secretion, or contraction of smooth or striated muscle (Fig. 4.2). The intermediary steps in this process include intracellular mobilization of calcium and multiple phosphorylation steps. Neurotransmitter release from neurons, adenosine diphosphate (ADP) release during platelet aggregation, and inflammatory mediator release from mast cells are widely regarded as secretory events involving the fusion of transmitter granules with the plasma membrane and extrusion by means of exocytosis.

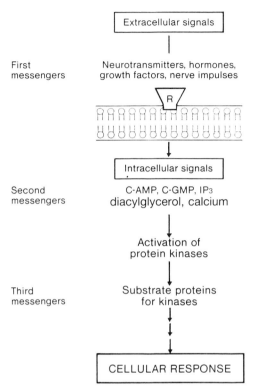

FIG. 4.2. Scheme showing sequence of events between extracellular receptor (R) activation and cellular response.

Ethanol in concentrations of 40–160 mM has been shown to depress the electrically-stimulated contractions of the guinea pig ileum longitudinal muscle/myenteric plexus preparation in a dose-dependent way (Mayer *et al.* 1980).

Similarly ethanol inhibits the spontaneous and acetylcholine-induced contractions of canine antral muscle (Sanders and Bauer 1982). Aggregation, induced by collagen, thrombin, and the calcium ionophere A23187, was inhibited by ethanol in physiologically tolerable concentrations (Haut and Cowan 1974; Fenn and Littleton 1982).

Calcium mobilization is a key link in the SSC chain. Interestingly, the inhibition by ethanol *in vitro* can be reversed by raising the extracellular calcium concentrations (Mayer *et al.* 1980). Ethanol has also been observed to inhibit the uptake of calcium in isolated synaptosomes from mouse and rat brain (Stokes and Harris 1982).

PHOSPHOINOSITIDES AND TRANSMEMBRANE SIGNALLING

A recent development in the understanding of the detailed events in SSC has come through the developments in the knowledge of the poly-phosphoinositides and their role in signal transduction (for reviews, see Berridge 1984; Berridge and Irvine 1984). Phosphoinositides are phosphorylated derivatives of the minor membrane phospholipid phosphatidyl inositol (PI). Interest has been focused on PI as possibly being linked with neurosecretion since the early observations by Hokin and Hokin (1953) that cholinergic stimulation of secretory glands led to a specific incorporation of ^{32}P in PI. This observation was linked by Michell (1975) to the mobilization of calcium. It is now realized that a number of hormones, growth factors, and neurotransmitters which use calcium as a second messenger specifically hydrolyse membrane phosphoinositides (Fig. 4.3). This may be part of a fundamental mechanism translating the message from the receptor activation across the membrane, to initiate a cascade of events resulting in the mobiliza-tion of calcium, activation of protein kinase C, and the release of arachidonic acid, the precursor of prostaglandins and leukotrienes (for review, see Oliw *et al.* 1983).

The key reaction is the hydrolysis of phosphatidyl inositol bis-phosphate (PIP$_2$) by the membrane phosphodiesterase (PDE) to give two products, namely inositol 4,5 trisphosphate (IP$_3$) and diacylglycerol (DG). IP$_3$ functions as an intracellular second messenger to mobilize

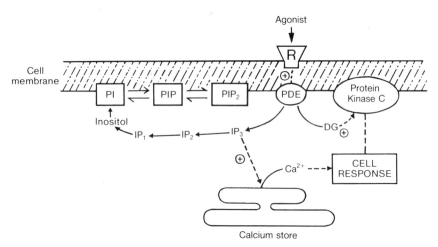

FIG. 4.3. Proposed model for phosphoinositides in receptor (R) activation. Hydrolysis of PIP$_2$ by the membrane phosphodiesterase (PDE) into inositol triphosphate (IP$_3$) and diacylglycerol (DG).

calcium. The calcium triggers subsequent steps in the signal cascade, dependent on calmodulin (CaM) kinase. IP_3 is rapidly removed by phosphatases hydrolysing it to inositol, 1,4 bisphosphate (IP_2) and inositol-1-phosphate (IP_1) which is returned to the inositol pool, for renewed synthesis of PI.

The other product of the hydrolysis of PIP_2 is diacylglycerol, DG. DG may have several functions. One might be to activate the kinase C (for review, see Nishizuka 1984) to phosphorylate specific proteins; another function could be to serve as a precursor of prostaglandins and leukotrienes through the release of arachidonic acid.

The receptor-mediated activation of PIP_2 hydrolysis thus represents a mechanism for a transmembrane signalling step in the SSC process. It seems to have a considerable degree of generality, as a great variety of agonists (such as acetylcholine, 5-hydroxytryptamine, histamine, peptides and vasopressin, and growth factors such as epidermal growth factor and platelet-derived growth factor) utilize this ubiquitous signalling system.

The key feature of the mechanism is the receptor-mediated activation of the phosphodiesterase (PDE, or phospholipase C) responsible for the cleavage of PIP_2 into IP_3 and DG. The level of PIP_2 is maintained by the activity of the PDE and the balance of the kinases and monoesterases that phosphorylate and remove the phosphate groups from the phosphoinositides. These enzymes are some of the most active in the cell. The mechanisms controlling the activity for PDE, and trigger enzyme in this signal cascade, are therefore of prime importance. It is possible that the PDE is controlled to some extent by the fluidity of the annular lipids surrounding it. If so, ethanol might be expected to interfere with one of the key elements in the SSC mechanism (cf. Fig. 4.1).

Ethanol and inositol phosphates

Experiments performed with R. Irvine at the Institute of Animal Physiology, Babraham, Cambridge, were stimulated by the observations by Allison and Cicero (1980) that injections of ethanol in the rat caused a dose-dependent reduction in the levels of IP in the cerebral cortex. A hypothesis was formulated that ethanol, which is a general depressant, could inhibit the neuronal events in the brain by interfering with PDE controlling the release of IP_3. To test this hypothesis, the following experiments were performed to study the effect of ethanol on polyphosphate inositides and inositol phosphates of the rat brain.

Male Sprague-Dawley rats were anaesthetized and ^{3}H-myo-inositol injected intracerebrally in a volume of 1–3 μl. The injection was made stereotactically 1 mm left of the bregma to a depth of 7 mm. After a period of 16–20 hours the animals were given either ethanol i.p. in a 25% (volume/volume) solution in distilled water or a corresponding volume of saline. The dose was 1, 2 and 4 g/kg. Thirty minutes later the rats were killed by cervical dislocation, decapitated, and the heads immediately immersed in liquid nitrogen; inositol phosphates and inositol lipids were extracted and analysed by anion exchange chromatography as described previously (Irvine *et al.* 1985).

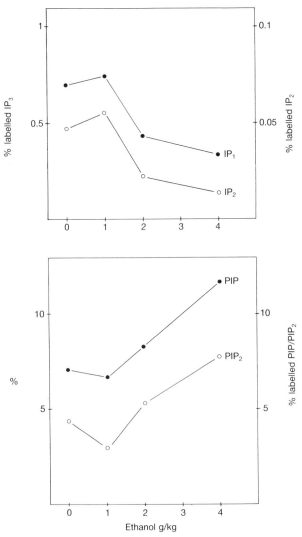

FIG. 4.4.   Effect of ethanol on levels of labelled inositolphosphate IP (top) and phosphoinositides PIP (bottom) in rat brain.

The effect of ethanol on the incorporation of ^{3}H-inositol in inositol lipids and inositol phosphates is shown in Fig. 4.4. At the lowest dose, 1 g/kg, when the rats were still awake, a nonsignificant elevation of IP$_1$ and IP$_2$ was seen. At 2 and 4 g/kg, when the rats had lost the righting reflex, the incorporation of ^{3}H-inositol in IP$_1$ and IP$_2$ was significantly depressed. The pattern of incorporation of ^{3}H-inositol in the phosphoinositides was the reverse. The fraction of PIP and PIP$_2$ as percentage of total lipid soluble radioactivity increased with the ethanol dose. The levels of ^{3}H-IP$_3$ were very low, and in some experiments not measureable. This is most likely due to the extremely rapid *post mortem* conversion of IP$_3$ to IP$_2$ and IP$_1$.

One possible explanation for the reduced levels of ^{3}H-IP$_3$ following ethanol treatment could be an inhibition of the metabolically-sensitive PI/PIP$_2$ cycle. This mechanism could, however, be excluded by the observation of elevated levels of ^{3}H-labelled PIP and PIP$_2$ after ethanol (Fig. 4.4, bottom). The most likely site of action is therefore on the phosphodiesterase (PDE), catalysing the cleavage of phosphoinositides into diacylglycerol and inositol phosphates.

Several other studies support these conclusions. Ethanol *in vitro* inhibited basal cerebral cortical PIP$_2$ breakdown in the mouse from 700 mM (Hoffman *et al.* 1986). Similar effects were seen in rat brain cortical slices (Gonzales *et al.* 1986). Ethanol intake by female rats during gestation and lactation prevented the metabolic degradation of the pups (Shah *et al.* 1984). In the mouse brain, carbachol-induced stimulation of PIP$_2$ breakdown was decreased following chronic intake of ethanol (Hoffman *et al.* 1986).

Together these studies provide convincing evidence of an action of ethanol on phosphoinositide breakdown, most likely on the phosphodiesterase. The likely mechanism is illustrated in Figure 4.5. These results confirm and extend those obtained by Allison and Cicero (1980). These workers observed an ethanol-induced depression of IP$_1$ in rat cortex, but were unable to draw conclusions on the mechanism involved. The experiments described above measured the degree of incorporation of ^{3}H-inositol in inositol lipids and IPs, and could thus exclude an effect of ethanol on the conversion of glucose-6-P to IP$_1$. In addition an effect of ethanol on IP$_2$, the immediate precursor to IP$_1$, was observed.

CONCLUSION

Cell membrane phosphoinositides have recently been suggested to have a key role in transmembrane signalling, i.e. the translation of extracellular receptor activation to intracellular events and resulting cellular response,

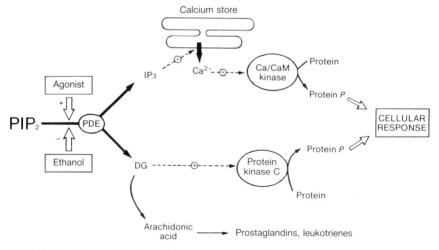

FIG. 4.5.   Scheme showing the proposed site of action of ethanol in phospho-inositide-dependent transmembrane signalling.

by the activation of a phosphodiesterase. Membrane phosphoinositides are cleaved into inositol trisphosphate and diacylglycerol, which serve as intracellular messengers mobilizing calcium and stimulating phospho-kinases. Several independent studies have now shown an inhibition of ethanol on phosphoinositide breakdown in mouse and rat brain. It is therefore suggested that ethanol, by fluidizing the cell membrane, alters the conformation of the phosphodiesterase, decreasing its activity, and thus inhibiting the generation of the intracellular messengers necessary for the final steps of the stimulus secretion coupling.

These observations make possible a new insight into the molecular mechanism of ethanol on the brain, and provide a unifying explanation of previous findings on increased fluidity of cell membranes and inhibition of SSC and neurotransmitter release.

REFERENCES

Allison, J. H. and Cicero, T. J. (1980). Alcohol acutely depresses myoinositol-1-levels in the male rat cerebral cortex. *J. Pharmacol. Exper. and Ther.* **213**, 24–7.

Berridge, M. J. (1984). Inositol trisphosphate and diacylglycerol as second messengers. *Biochem. J.* **220**, 345–60.

——— and Irvine, R. F. (1984). Inositol trisphosphate. A novel second messenger. *Nature (London)* **312**, 315–21.

Carmichael, F. J. and Israel, Y. (1975). Effects of ethanol on neurotransmitter release by rat brain cortical slices. *J. Pharmacol. Exper. and Ther.* **193**, 824–34.

Chin, J. H. and Goldstein, D. B. (1977). Effects of low concentrations of ethanol of the fluidity of spin labelled erythrocyte and brain membranes. *Mol. Pharmacol.* **13**, 435–41.

Fenn, C. G. and Littleton, J. M. (1982). Inhibition of platelet aggregation by ethanol *in vitro* shows specificity for aggregating agent used and is influenced by platelet lipid composition. *Thromb. Haemostas. (Stuttgart)* **48**, 49–53.

Gonzales, R. A. *et al.* (1986). Effects of ethanol on stimulated inositol phospholipid hydrolysis in rat brain. *J. Pharmacol. Exper. and Ther.* **237**, 92–9.

Haut, M. and Cowan, D. (1974). The effect of ethanol on haemostatic properties of human blood platelets. *Am. J. Med.* **56**, 22–33.

Hoffman, P. L. *et al.* (1986). Acute and chronic effects of ethanol on receptor-mediated phosphatidylinositol 4,5-bisphosphate breakdown in mouse brain. *Mol. Pharmacol.* **30**, 13–18.

Hokin, M. R. and Hokin, L. E. (1953). Enzyme secretion and the incorporation of P^{32} into phospholipids of pancreas slices. *J. Biol. Chem.* **203**, 967–77.

Irvine, R. F., Anggard, E., and Downes, P. C. (1985). Metabolism of inositol (1,4,5) triphosphate and inositol (1,3,4) triphosphate in rat parotid glands. *Biochem. J.* **229**, 505–11.

Kalant, H. and Grose, W. (1967). Effects of ethanol and pentobarbital on release of acetylcholine from cerebral cortex slices. *J. Pharmacol. Exper. and Ther.* **158**, 386–93.

Mayer, J. M., Khanna, J. T., and Kalant, H. (1980). A role for calcium in the acute and chronic actions of ethanol *in vitro*. *Europ. J. Pharmacol.* **68**, 223–7.

Michell, R. H. (1975). Inositol phospholipids and cell surface receptor function. *Biochim. Biophys. Acta* **415**, 81–147.

Nishizuka, Y. (1984). Turnover of inositol phospholipids and signal transduction. *Science* **225**, 1365–70.

Oliw, E., Granstrom, E., and Anggard, E. (1983). The prostaglandins and essential fatty acids. In *Prostaglandins and related substances* (eds. C. R. Pace-Asciak and E. Granstrom). Elsevier, Amsterdam.

Sanders, K. M. and Bauer, A. J. (1982). Ethyl alcohol interferes with excitation—contraction, mechanisms of canine antral muscle. *Am. J. Physiol.* **242**, G222–30.

Shah, I. R. *et al.* (1984). Effect of ethanol on rat brain polyphosphoinositides. *J. Neurochem.* **42**, 873–4.

Stokes, J. A. and Harris, R. A. (1982). Alcohols and synaptosomal calcium transport. *Mol. Pharmacol.* **22**, 99–104.

Taraschi, T. F. and Rubin, E. (1985). Effects of ethanol on the chemical and structural properties of biologic membranes. *Laboratory Invest.* **52**, 120–31.

5

Adaptation in neuronal Ca^{2+}-channels may cause alcohol physical dependence

JOHN LITTLETON, JOYCE HARPER, MIKE HUDSPITH,
CHRISTOS PAGONIS, SIMON DOLIN, and
HILARY LITTLE

INTRODUCTION

Ethanol *in vitro* inhibits the depolarization-induced entry of Ca^{2+} into nerve terminals (Leslie *et al.* 1983) and, probably in consequence, inhibits depolarization-induced release of neurotransmitters (Lynch and Littleton 1983). When preparations of brain are taken from animals which have been made tolerant and physically dependent on ethanol, Ca^{2+} entry into nerve terminals is reduced (Leslie *et al.* 1983) but the release of transmitters is either normal or enhanced (Clark *et al.* 1977; Lynch and Littleton 1983). This implies that some increase in sensitivity to Ca^{2+} must have occurred in central neurones in response to the continued presence of ethanol in the brain (Lynch and Littleton 1983). This change could be considered as an adaptive mechanism to the inhibitory effect of ethanol, and is a potential cause of physical dependence since, on removal of ethanol, neuronal hyperexcitability resulting from increased Ca^{2+} sensitivity could cause the withdrawal syndrome.

Although the total Ca^{2+} entry into nerve terminals induced by depolarization is reduced in preparations from ethanol-dependent animals (Leslie *et al.* 1983), this does not mean that Ca^{2+} flux through all types of voltage-operated Ca^{2+} channels is necessarily reduced. Electrophysiological evidence suggests at least three types of neuronal Ca^{2+} channel (Nowycky *et al.* 1985) one of which is sensitive to drugs of the Ca^{2+} 'antagonist' group (especially dihydropyridines, DHP). Recent evidence from cell cultures suggests that the major site for the dihydrophyridine-sensitive Ca^{2+} channels may be on the nerve cell body, whereas Ca^{2+} channels on nerve terminals may be mainly DHP-insensitive (Kongsamut and Miller 1986). It is possible therefore that increased voltage-operated Ca^{2+} entry into the soma of nerves might explain the apparent increase in neuronal Ca^{2+} sensitivity without this causing any increase in Ca^{2+} entry into nerve terminal preparations. Ca^{2+} entry into the soma causes transmitter release from this part of the nerve cell (Suetake *et al.* 1981) and might modulate the sensitivity of the release process from terminals in intact neurones, a possible mechanism being the inositol lipid signalling system (Tanaka *et al.* 1986). There is good evidence that increased activity in this system can increase the Ca^{2+} sensitivity of neurotransmitter release, and that it can be influenced by drugs of the dihydropyridine type (Kendal and Nahorski 1985).

The working hypothesis before beginning the work described here can therefore be stated as: The presence of ethanol in brain evokes an adaptive increase in the dihydropyridine-sensitive sub-type of Ca^{2+} channels. This increases the sensitivity of central neurones to Ca^{2+} by modulation of their inositol lipid signalling system rather than by causing a significant increase in the total depolarization-induced Ca^{2+} entry into nerve terminals.

EVIDENCE FOR INCREASED DHP-SENSITIVE Ca^{2+} CHANNELS

The number of Ca^{2+} channels sensitive to DHP drugs can be simply estimated using standard radioligand binding techniques. The ligands used most commonly are [³H] nitrendipine and [³H]nimodipine, both DHP Ca^{2+} 'antagonists'. The first report of an alteration in the binding sites for these ligands associated with alcohol administration is that of Ross and Monis (1984). Although as yet only in abstract form, this is an interesting report in that it shows a *c.* 60 per cent increase in [³H]DHP binding sites in mouse brain after a large acute dose of ethanol. This increase was observed after ethanol administration to a strain of mice (C57/BL) which develops tolerance very rapidly, but not in a strain (DBA/2) which develops tolerance slowly (Grieve and Littleton 1979).

Similar findings in rats given ethanol acutely by injection were reported by Lucchi *et al.* (1985); these authors showed an increase in [³H]DHP binding sites of a magnitude similar to that of Ross and Monis. After chronic administration of ethanol in drinking water, however, they reported that [³H]DHP binding sites in brain were no longer increased, and even showed a small decrease. The regime of ethanol administration used in these experiments seems very unlikely to lead to the maintained high concentrations of ethanol in brain required to evoke tolerance and physical dependence.

We have investigated the effect of chronic administration of ethanol to laboratory rats by inhalation, using a regime which has consistently been shown to produce tolerance and physical dependence (Lynch and Littleton 1983). The binding of [³H]nimodipine to cerebral cortical membranes from these animals was investigated using a modification of the method of Glossman and Ferry (1985). Representative results are shown in Fig. 5.1. The affinity of [³H]DHP for binding sites on cortical membranes is unchanged in brains from ethanol-dependent animals, but the number of binding sites is increased by 50 per cent. This increase is statistically significant at the $p < 0.05$ level.

There is recent evidence that similar adaptive changes in DHP-sensitive Ca^{2+} channels may be evoked in cell cultures of neuronal origin by growing these in media containing ethanol. Messing *et al.* (1986) using a rat phaeochromocytoma-derived cell line (PC12 cultures) showed a *c.* 100 per cent increase in DHP binding sites associated with growth for six days in the presence of ethanol, and this was associated with an increase in depolarization-induced entry of $^{45}Ca^{2+}$ into these cells. Cultures of this kind resemble neuronal cell bodies before differentiation and extension of neurites and nerve terminals (Kongsamut and Miller 1986). They have mainly DHP-sensitive Ca^{2+} channels on their membranes and thus

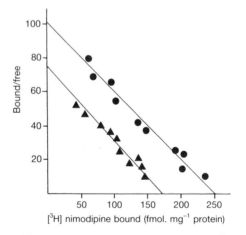

FIG. 5.1.   Effect of induction of ethanol dependence on [³H]nimodipine binding to rat cortical membranes. Five paired Scatchard analyses of [³H]nimodipine binding were performed on cortical membranes from control (▲) and ethanol-dependent (●) rats. The figure shows a typical result. For control animals, the number of binding sites (B_{max}) was 131 ± 14 fmol·mg⁻¹ protein; for ethanol-dependent animals, the value was 196 ± 13 fmol·mg⁻¹ protein. The increase in binding sites in the ethanol-dependent animals is significant at the $p < 0.05$ level. The binding affinity (K_D) is similar in control (0.50 ± 0.05 nM) and in ethanol-dependent animals (0.59 ± 0.05 nM).

provide a good model for studying adaptation in these. However, they are tumour cells and their propensity for growth and division makes their appropriateness as a model for neuronal adaptation doubtful. Using primary cultures of bovine adrenal chromaffin cells which do not suffer from this disadvantage, functional effects in ethanol-grown cultures have been obtained, which suggest that similar increases in DHP-sensitive Ca^{2+} channels occur in these (see below).

EVIDENCE FOR INCREASED DHP-SENSITIVE INOSITOL LIPID BREAKDOWN

As yet, very little work has been published on the effect of ethanol administration on the inositol lipid signalling system in brain. Most work on the system in general has concentrated on its role in receptor-mediated cellular responses, and two papers refer to alterations in agonist-induced changes in phosphatidyl inositol (PI) metabolism in brain after chronic alcohol administration. The first report used ^{32}P incorporation as an index of PI breakdown, and reported only small alterations in receptor-mediated PI breakdown after ethanol (Smith 1983). The second used [^3H]inositol incorporation and accumulation of [^3H]inositol phosphates as the index, and reported a significant enhancement of PI breakdown initiated by the muscarinic receptor agonist, carbachol, in cortices of ethanol-dependent mice (Hoffman *et al.* 1986). This correlated with an increased number of muscarinic binding sites in these preparations.

We have investigated the depolarization-induced breakdown of PI in brain preparations for some time, using methods similar to those described above. This is a much neglected area because it is currently impossible to prove that such breakdown is not simply an indirect consequence of activation of receptors by transmitters released by the depolarization. All the evidence which we (Hudspith *et al.* 1987) and others (Batty *et al.* 1985) have obtained is equivocal, but suggests that part of the depolarization-induced PI breakdown is not a consequence of transmitter release, and that this fraction of PI breakdown is more sensitive to drugs of the DHP type than is the release of any conventional neurotransmitter. We believe therefore, that DHP-sensitive Ca^{2+} channels are, in some way, directly coupled to the inositol lipid signalling system and that via this system, they may modulate both receptor-mediated PI breakdown (Fowler *et al.* 1986) and neurotransmitter release (Tanaka *et al.* 1986).

The chronic administration of ethanol to rats by inhalation to induce tolerance and dependence is associated with a marked increase in the depolarization-induced breakdown of PI in cortical slice preparations (Hudspith *et al.* 1985). The relationship of this increase to the presence of DHP-sensitive Ca^{2+} channels can be assessed by the use of the Ca^{2+} 'agonist', BAY K 8644. This compound acts as a partial activator of DHP-sensitive Ca^{2+} channels; in other words it prolongs the open-state of the channel. The effect of BAY K 8644 on the K^+ depolarization-induced breakdown of PI in cortical preparations from control and ethanol-dependent rats is shown in Figure 5.2. Under these conditions, BAY K 8644 has very little enhancing effect on inositol phospholipid breakdown in control preparations, but a marked and highly significant potentiating effect in preparations from ethanol-dependent rats. The presence of the Ca^{2+} 'antagonist', nitrendipine, completely prevents this potentiation at concentrations which alone have no effect on PI breakdown. The conditions used in these experiments were such that the dihydropyridine drugs had minimal effects on transmitter release (see below) suggesting that the changes observed were not simply an indirect

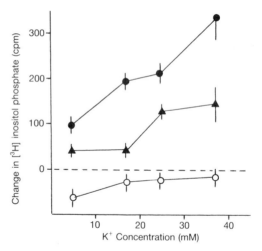

FIG. 5.2.   Effect of dihydropyridine drugs on inositol phospholipid breakdown in rat cortical slices. The figure shows the change in accumulation of [³H]-inositol phosphates produced by incubating cortical slices with dihydropyridines at different K^+ concentrations. The values obtained with K^+ depolarization alone have been subtracted. The Ca^{2+} channel activator BAY K 8644 at 5×10^{-7} M has little effect on inositol phosphate accumulation from control animals (▲) but produces a greater and highly significant stimulation in preparations from ethanol-dependent rats (●). This stimulation is completely prevented by the concomitant presence of the Ca^{2+} channel inhibitor, nitrendipine (○) at a concentration of 10^{-5} M. All values are means $\pm$ s.e.m. of at least five determinations.

consequence of receptor activation by released transmitters. They could therefore reflect direct coupling between an increased number of DHP-sensitive Ca^{2+} channels and inositol lipid signalling in ethanol-dependent brain preparations.

EVIDENCE FOR INCREASED DHP-SENSITIVE NEUROTRANSMITTER RELEASE

In general, in using neuronal preparations it has proved difficult to demonstrate any effect of the DHP drugs on depolarization-induced release of neurotransmitters. However, using the DHP Ca^{2+} channel activator, BAY K 8644, under conditions of partial K^+ depolarization, increased release of 5-hydroxytryptamine (5-HT) (Middlemiss and Spedding 1984), acetylcholine (ACh) and noradrenaline (NA) (Middlemiss 1985) from brain preparations has been reported. We attempted to use similar methods to study the effect of BAY K 8644 on K^+-induced dopamine (DA) release from rat striatal slices and electrically-induced release of NA from rat cortical slices, with no success. At about this time Turner and Goldin (1985) reported that the DHP drugs only affected K^+ depolarization-induced $^{45}Ca^{2+}$ entry into brain preparations if the depolarizing stimulus was presented in buffers containing low Na^+ concentrations. They favoured the explanation that, under these conditions, the Na^+/Ca^{2+} exchange mechanism would be of less importance, allowing voltage-operated Ca^{2+} channels to make a proportionately greater contribution to Ca^{2+} flux.

In subsequent experiments on striatal slices, the release of $[^3H]DA$ induced by combinations of K^+-depolarization and BAY K 8644 in low Na^+ buffers of composition similar to those used by Turner and Goldin (1985) were followed; the results are shown in Figure 5.3. The potentiation of DA release caused by BAY K 8644 in striatal slice preparations from control rats does not reach the $p < 0.05$ level of significance, but the potentiation of release induced in preparations from ethanol-dependent animals is markedly greater, and easily reaches significance. The potentiation is partly inhibited by nitrendipine at a low concentration, which alone has no effect. Figure 5.4 shows similar experiments in which the effect of BAY K 8644 on electrically-stimulated release of $[^3H]NA$ from cortical preparations was studied. In these experiments, Na^+ flux was inhibited by the presence of the Na^+-channel inhibitor, tetrodotoxin. The results are essentially the same: BAY K 8644 potentiates release to a greater extent in the preparations from ethanol-dependent rats, and this potentiation is inhibited by nitrendipine.

An alternative explanation for the effects of low Na^+ buffers and

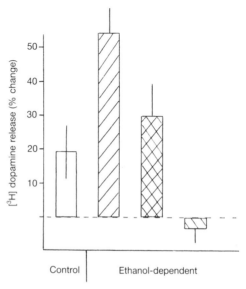

FIG. 5.3.   Effect of dihydropyridines on [³H]dopamine release from rat striatal slices. The figure shows the percentage change in fractional dopamine release produced by superfusing striatal slice preparations in low Na⁺ buffers with dihydropyridines while depolarizing with 25 mM K⁺. The effect of the K⁺ alone on dopamine release has been subtracted. In control preparations (extreme left), BAY K 8644 (5×10^{-7} M) produced a small non-significant increase in dopamine release. In preparations from ethanol-dependent animals (second column), BAY K 8644 (5×10^{-7} M) produced a marked and highly significant ($p < 0.01$) increase. Nitrendipine (5×10^{-6} M) inhibited the effect of BAY K 8644 (third column), but had no effect alone (extreme right). Values are means ±s.e.m. of at least five determinations.

tetrodotoxin in exposing the action of BAY K 8644 on transmitter release is possible. In both cases the role of neuronal conduction in activating the transmitter release system in the slice preparations would be inhibited. It may be that these circumstances favour transmitter release from nerve cell bodies in the preparation, rather than from nerve terminals. Since the cell bodies are thought to be a major site of DHP-sensitive Ca²⁺ channels, release from this site would be more sensitive to BAY K 8644 than that from nerve terminals.

If this is so, then adrenal chromaffin cells (which resemble undifferentiated neurones, and in which the catecholamine release process is highly sensitive to DHP drugs) in culture should provide a good model in which to investigate the reason for this increased functional activity of DHP-sensitive Ca²⁺ channels. Preliminary results from adrenal cell cultures grown in the presence of 200 mM ethanol have been obtained;

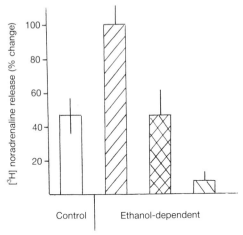

FIG. 5.4. Effect of dihydropyridines on [³H]noradrenaline release from rat cortical slices. The same conventions are followed as in Figure 5.3. In these experiments, release was induced by electrical stimulation (1 Hz, 100 mA, 2 msec, 15 min) in the presence of tetrodotoxin (5×10^{-8} M). Release obtained by this procedure alone has been subtracted from that obtained when dihydropyridines were present in the superfusing fluid. The increase in noradrenaline release produced by BAY K 8644 is significantly greater in preparations from ethanol-dependent animals than from controls ($p < 0.05$) and is inhibited by nitrendipine. Values are means ± s.e.m. of at least five determinations.

they suggest that these cells do indeed show an increase in functional activity of these channels, but they also suggest another, completely unexpected, mechanism. The results are shown in Figure 5.5. Control cultures, grown in the absence of ethanol, showed exactly the expected response, in that nitrendipine produced a concentration-related inhibition of catecholamine release, whereas BAY K 8644 produced a concentration-related potentiation of release. Cultures grown in the presence of ethanol showed enhanced depolarization-induced release of catecholamines which could be inhibited by nitrendipine, a result entirely consistent with an increase in functionally active DHP-sensitive Ca^{2+} channels on these cells. The effect of BAY K 8644 was, however, completely unexpected; at low concentrations this partial Ca^{2+} 'agonist' reduced catecholamine release from ethanol-grown cultures to a point where release was similar to that from control cultures. Thereafter BAY K 8644 enhanced catecholamine release with a concentration dependence identical to that in control cultures.

It must be stressed that these results are preliminary, and are open to several interpretations. The most likely however is that BAY K 8644 in ethanol-grown cultures is displacing a more potent endogenous agonist

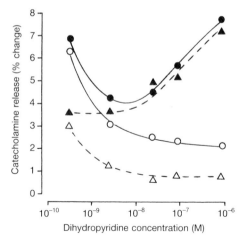

FIG. 5.5.   Effects of dihydropyridines and 18 mM K$^+$ on release of catechol-amines from adrenal cell cultures. The figure shows the percentage change from basal catecholamine release induced by incubating adrenal cell cultures with BAY K 8644 (filled symbols) or nitrendipine (open symbols). Control cultures (▲, △, dashed lines) were grown in normal medium for six days. Ethanol-grown cultures (●, ○, solid lines) were grown in medium containing 200 mM ethanol. Results are means of at least five experiments, error bars are omitted for clarity; a similar pattern was seen in each experiment.

from DHP binding sites at low concentrations, so that its net effect is to reduce the number of Ca^{2+} channels in the open-state. When the endogenous agonist is displaced, the activating effect of BAY K 8644 itself on the DHP-sensitive Ca^{2+} channels can be observed; the net effect is to increase the number of channels in the open state. If this interpretation is correct, then chronic ethanol treatment must enhance the production of a compound (transmitter ?) in cultures which stabilizes DHP-sensitive Ca^{2+} channels in the open state. What this compound may be, and whether similar effects can explain the changes observed in brain, await further experimentation.

EVIDENCE THAT INCREASED DHP-SENSITIVE Ca^{2+} CHANNELS CAUSE DEPENDENCE

If the increased neuronal Ca^{2+} sensitivity previously reported (Lynch and Littleton 1983) is an important factor in the latent hyperexcitability of the central nervous system which underlies physical dependence, and if the increased Ca^{2+} sensitivity is due to increased activity of DHP-sensitive Ca^{2+} channels, then two important predictions follow. First, potentiation of the activity of DHP-sensitive Ca^{2+} channels by administration of BAY K 8644 *in vivo* to control animals should mimic the ethanol physical withdrawal syndrome. Second, inhibition of the activity of DHP-sensitive Ca^{2+} channels by administration of Ca^{2+} 'antagonists' such as nitrendipine and nimodipine to ethanol-dependent animals should prevent the withdrawal syndrome. Both these predictions are supported by results.

The upper section of Figure 5.6 shows the effect of dihydropyridine drugs given to mice of the TO Swiss strain. Mice of this strain show a mild type of tonic-clonic convulsion when suspended by the tail. In control animals each convulsive episode is very brief and the phenomenon rapidly disappears as the animals become used to being handled in this way. In ethanol withdrawal, however, each episode is more

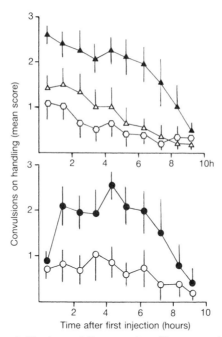

FIG. 5.6. Effects of dihydropyridines on handling convulsions in mice. The upper part of the figure shows the effect of two intraperitoneal injections (at 0 hours and 3 hours) of BAY K 8644 2 mg·kg^{-1} (▲), nimodipine 50 mg·kg^{-1} (△), or vehicle, 20% DMSO (◇) on the incidence of convulsions on handling in male TO Swiss mice. The lower section shows the effect of nimodipine, 12.5 mg·kg^{-1} (○) injected at the same time intervals on mice with handling convulsions undergoing ethanol withdrawal (from time 0). Injection of the vehicle (●) has no effect on the withdrawal convulsions. Each value is the mean ± s.e.m. convulsion score from at least eight animals.

prolonged and the phenomenon lasts for the *c.* 10 hour period which characterizes the withdrawal syndrome. The figure shows that the Ca^{2+}-channel activator, BAY K 8644, markedly potentiates the severity of these convulsions (assessed by the method of Goldstein and Pal 1971), whereas the Ca^{2+} channel inhibitor, nimodipine, has no effect on the convulsive state in control animals. In contrast, and confirming the results Little *et al.* (1986) for rats, nimodipine almost completely prevents the incidence of handling convulsions in ethanol-withdrawn mice (lower section of Fig. 5.6). These results are entirely consistent with the hypothesis that ethanol physical dependence is caused by an increase in number and functional activity of DHP-sensitive Ca^{2+} channels on central neurones. They suggest that, in normal animals, these channels play little part in determining central nervous system excitability but that, as an adaptive response to the depressant action of ethanol, they induce a state of latent hyperexcitability in brain which, if exposed, leads to the physical withdrawal syndrome.

CONCLUSIONS

The results described provide strong circumstantial evidence for the original hypothesis that the presence of ethanol in brain evokes an adaptive increase in the dihydropyridine-sensitive sub-type of Ca^{2+} channels. The development of ethanol tolerance and physical dependence is associated with an increase in the number of DHP-binding sites on rat cortical membranes, as assessed by radioligand binding techniques. That these binding sites represent functional DHP-sensitive Ca^{2+} channels is strongly suggested by the enhanced ability of the Ca^{2+} channel activator, BAY K 8644, to stimulate inositol lipid breakdown in cortical slices from ethanol-dependent rats, and to stimulate the release of [³H]noradrenaline (cortical slices) and [³H]dopamine (striatal slices). The ability of Ca^{2+} 'antagonists' of the DHP type to inhibit these effects, and to prevent the physical syndrome of withdrawal from ethanol, strengthens the suggestion that the increase in number and activity of these Ca^{2+} channels is the cause of ethanol physical dependence. Whether some endogenous ligand of the DHP-sensitive Ca^{2+} channel is involved, as suggested by the preliminary cell culture experiments, remains an exciting possibility.

These experiments at last offer the possibility of a logical therapeutic intervention into the processes of development of alcohol tolerance, physical dependence, and withdrawal. The DHP Ca^{2+} 'antagonist' drugs do not merely substitute for the effect of ethanol; instead they interfere directly with the adaptive process which has instituted tolerance and dependence. They should therefore have minimal effects on normal subjects, and should have little abuse potential of their own. To date these drugs are not particularly effective on the central nervous system, but a new generation of compounds, designed with these effects in mind, provides a very bright prospect for the treatment of alcoholism and, indeed, dependence on other central depressant drugs.

ACKNOWLEDGEMENTS

Much of this work was supported by a grant from the Wellcome Trust. Simon Dolin holds a Saw Medical Fellowship from the University of Western Australia. Drugs were provided by Bayer, UK.

REFERENCES

Batty, I., Kendal, D. A., and Nahorski, S. R. (1985). Endogenous acetylcholine release mediates inositol phospholipid hydrolysis induced by depolarising stimuli in cortical slices. *Brit. J. Pharmacol.* **8**, 108P.

Clark, J. W., Kalant, H., and Carmichael, F. J. (1977). Effect of ethanol tolerance on release of acetylcholine and norepinephrine by rat cerebral cortical slices. *Canadian J. Physiol. and Pharmacol.* **55**, 758–68.

Fowler, C. J., O'Carroll, A. M., Court, J. A., and Candy, J. M. (1986). Stimulation by noradrenaline of inositol phospholipid breakdown in the rat hippocampus: effect of the ambient potassium concentration. *J. Pharm. and Pharmacol.* **38**, 201–8.

Glossman, H. and Ferry, D. R. (1985). Assay for calcium channels. *Methods in Enzymology* **109**, 513–50.

Goldstein, D. B. and Pal, N. (1971). Alcohol dependence produced in mice by inhalation of ethanol: grading the withdrawal reaction. *Science* **172**, 288–90.

Grieve, S. J. and Littleton, J. M. (1979). Age and strain differences in the rate of development of tolerance to ethanol by mice. *J. Pharm. and Pharmacol.* **31**, 696–701.

Hoffman, P. L., Moses, F., Luthin, G. R., and Tabakoff, B. (1986). Acute and chronic effects of ethanol on receptor-mediated phosphatidylinositol 4,5,-bisphosphate breakdown in mouse brain. *Mol. Pharmacol.* **30**, 13–18.

Hudspith, M. J., John, G. R., Nhamburo, P. T., and Littleton, J. M. (1985). Effects of ethanol *in vitro* and *in vivo* on calcium-activated metabolism of membrane phospholipids in rat synaptosomal and brain slice preparations. *Alcohol* **2**, 133–8.

——, Brennan, C. H., Charles, S., and Littleton, J. M. (1987). Dihydropyridine-sensitive Ca²⁺ channels and inositol phospholipid metabolism in ethanol physical dependence. *Ann. New York Acad. of Sciences* **492**, 156–70.

Kendal, D. A. and Nahorski, S. R. (1985). Dihydropyridine calcium channel activators and antagonists influence depolarization-evoked inositol phospholipid hydrolysis in brain. *European J. Pharmacol.* **115**, 31–6.

Kongsamut, S. and Miller, R. J. (1986). Nerve growth factor modulates the drug sensitivity of neurotransmitter release from PC-12 cells. *Proc. Nat. Acad. Sciences of the USA* **83**, 2243–7.

Leslie, S. W., Barr, E., Judson, C., and Farrah, R. P. (1983). Inhibition of fast- and slow-phase depolarisation-dependent synaptosomal calcium uptake by ethanol. *J. Pharmacol. and Exper. Ther.* **225**, 571–5.

Little, H. J., Dolin, S. J., and Halsey, M. J. (1986). Calcium channel antagonists decrease the ethanol withdrawal syndrome in rats. *Life Sciences* **39**, 2059–65.

Lucchi, L., Govoni, S., Battaini, F., Pasinetti, G., and Trabucchi, M. (1985). Ethanol administration *in vivo* alters calcium ions control in rat striatum. *Brain Research* **332**, 376–9.

Lynch, M. A. and Littleton, J. M. (1983). Possible association of alcohol tolerance with increased synaptic Ca^{2+} sensitivity. *Nature* **305**, 175–6.

Messing, R. O., Carpenter, C. L., Diamond, I. and Greenberg, D. A. (1986). Ethanol regulates calcium channels in clonal neural cells. *Proc. Nat. Acad. of Sciences of the USA* **83**, 6213–15.

Middlemiss, D. N. (1985). The calcium channel activator, BAY K 8644, enhances K^+-evoked efflux of acetylcholine and noradrenaline from rat brain. *Naunyn Schmiederberg's Arch. of Pharmacol.* **331**, 114–16.

—— and Spedding, M. (1984). A functional correlate for the dihydropyridine binding site in rat brain. *Nature* **303**, 175–6.

Nowycky, M. C., Fox, A. P., and Tsien, R. W. (1985). Three types of neuronal Ca^{2+} channel with different calcium agonist sensitivity. *Nature* **316**, 440–3.

Ross, D. H. and Monis, N. (1984). [^{3}H]-nitrendipine binding in alcohol-preferring mice. *Alcoholism: Clin. and Exper. Research* **8**, 88.

Smith, T. L. (1983). Influence of chronic ethanol consumption on muscarinic cholinergic receptors and their linkage to phospholipid metabolism in mouse synaptosomes. *Neuropharmacol.* **22**, 661–3.

Suetake, K., Kojima, H., Inaga, K., and Koketsu, K. (1981). Release of transmitters from nerve terminals and soma. *Brain Research* **205**, 436–40.

Tanaka, C., Fujiwara, H., and Fuhii, Y. (1986). Acetylcholine release from guinea pig caudate slices evoked by phorbol ester and calcium. *Fed. European Societies Letters* **195**, 129–34.

Turner, T. J. and Goldin, S. M. (1985). Calcium channels in rat brain synaptosomes: identification and pharmacological characterisation. High affinity blockade by organic Ca^{2+} channel blockers. *J. Neuroscience* **5**, 841–9.

6

Drug anticipation and drug tolerance*

SHEPARD SIEGEL

Most theories of drug tolerance emphasize the physiological consequences of repeated pharmacological stimulation; they postulate some systemic change within the organism as a result of early drug administrations that either modifies receptor sensitivity to the drug, induces neurochemical changes, or alters the metabolism of the drug. It has become apparent, however, that a complete account of tolerance requires an appreciation of environmental influences.

ENVIRONMENTAL SPECIFICITY OF TOLERANCE

The importance of environmental influences on tolerance is illustrated by the results of a number of studies demonstrating that tolerance is not the inevitable result of repeated drug administration. Rather, the drug-experienced organism may or may not display tolerance, depending upon whether the drug is administered in the usual drug administration environment or an alternative environment.

Environmental specificity of morphine tolerance

Early demonstrations of the contribution of pre-drug cues to tolerance were presented by Mitchell and colleagues in a series of 12 studies (e.g. Adams *et al.* 1969; Siegel 1978*a*). In these experiments, rats responded in the expected analgesia-tolerant manner to the last of a series of morphine injections only if this final injection occurred in the same environment as the prior injections of morphine. Much subsequent research has extended these demonstrations of the importance of pre-drug cues in tolerance to the analgesic effect of morphine (Siegel and MacRae 1984).

*Based on the Dent Memorial Lecture.

The details of the designs of experiments demonstrating the environmental specificity of morphine analgesic tolerance differed, but all incorporated two groups of rats, both receiving the drug a sufficient number of times for tolerance to develop during the initial, tolerance-development phase of the experiment. The analgesic effect of the drug was evaluated in a subsequent tolerance-test phase. For one of the two groups, this test was conducted following the same cues that signalled the drug during the tolerance development phase (Same-Tested). For the second group, the tolerance test was conducted following different cues from those that signalled the drug during the tolerance development phase (Different-Tested). In addition, the design of the experiments enabled evaluation of the magnitude of the analgesic response in rats receiving the drug for the first time (Control). Results obtained during the tolerance test in a number of experiments using this procedure are summarized in Figure 6.1.

Figure 6.1A summarizes results of an experiment in which analgesia in rats was assessed following a tenth morphine injection (5 mg/kg), this tenth injection being paired with an audiovisual cue (Siegel *et al.* 1978). Pain sensitivity was measured by noting the rat's latency to lick a paw when placed on a 54°C surface (the 'hot plate' procedure). For Same-Tested rats, the nine pretest injections were signalled by the same cue that signalled the test injection. Different-Tested rats, in contrast, received their nine prior drug injections and cue presentations in an unpaired manner. As can be seen by comparing Control rats with Same-Tested rats, tolerance to the analgesic effect of morphine was obtained; Control rats, which received the drug for the first time on the tolerance-test session, were significantly less sensitive to the thermal stimulation (i.e., were more analgesic) than Same-Tested rats, which received the drug for the tenth time on the test session. However, results obtained from Different-Tested rats demonstrate that analgesic tolerance may not be displayed following repeated morphine administration. Different-Tested rats had the same pharmacological history as Same-Tested rats (i.e., they received the same dose of morphine, equally often, and at the same intervals), but Different-Tested rats were as profoundly analgesic as Control animals. Other studies, using different drug doses and/or analgesia assessment procedures, have similarly demonstrated that Same-Tested rats are more tolerant than Different-Tested rats, although such situational-specificity of tolerance is not always complete; that is, both groups of drug-experienced animals may be more tolerant than drug-naïve Control animals.

Figure 6.1B illustrates results of an experiment in which pain sensitivity was assessed with a paw-pressure analgesiometer; the rat was free to withdraw its paw from a source of gradually and constantly

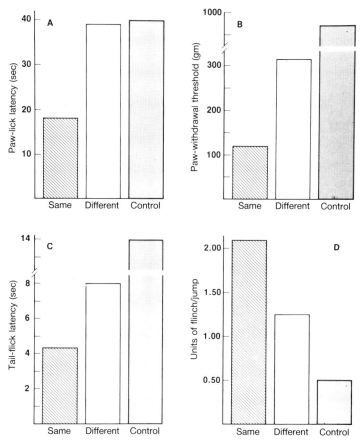

FIG. 6.1. Environmental-specificity of tolerance to the analgesic effect of morphine in the rat, demonstrated using the hot plate, paw-lick (A), paw-pressure (B), tail-flick (C), and flinch/jump (D) analgesia assessment procedures. These results are based on Siegel *et al.* (1978), Siegel (1976), Advokat (1980), and Tiffany and Baker (1981), respectively.

increasing pressure, with the amount of pressure applied before the withdrawal response occurs (i.e., the paw-withdrawal threshold) providing a measure of pain sensitivity (Siegel 1976). As can be seen, Same-Tested rats, with a pre-test history of eight morphine injections (5 mg/kg), evidenced greater sensitivity to the paw pressure (i.e., more analgesic tolerance) than the equally drug-experienced Different-Tested rats. Thus, again, equivalent opiate exposure does not lead to equivalent levels of tolerance. As is apparent in Figure 6.1B, both Same- and Different-Tested rats displayed more tolerance than Control rats, indicating a non-situational component of tolerance as well.

A similar pattern of results may be seen (Advokat 1980) in a study of analgesic tolerance, as assessed by tail-flick latency, following eight injections of morphine (7.5 mg/kg). As depicted in Figure 6.1C, Control rats are profoundly analgesic. Although both drug-experienced groups display shorter tail-flick latencies than the Control level, they differed significantly; Same-Tested animals were more tolerant than Different-Tested animals. This pattern of results was confirmed in a different experiment (Tiffany and Baker 1981), using a different number of pre-test sessions (five), a different dose of morphine (20 mg/kg), and a different analgesia-assessment procedure (digitalized flinch/jump magnitude to electric shocks, with smaller numbers indicating less sensitivity to shock, i.e., greater analgesia). The results of this experiment are presented in Figure 6.1D. Again, Same-Tested animals were more tolerant than Different-Tested animals, although neither was as analgesic as Control animals.

The environmental-specificity of tolerance to the analgesic effect of morphine is rather general, having recently been demonstrated in the terrestrial gastropod snail, *Cepaea nemoralis* (Kavaliers and Hirst 1986). Such findings suggest that such specificity of tolerance involving classical conditioning 'may be a general phenomenon having an early evolutionary development and broad phylogenetic continuity' (Kavaliers and Hirst 1986, p. 1201).

The finding that morphine tolerance is more pronounced in the drug administration environment than an alternative environment has been demonstrated with respect to effects of the drug other than analgesia. It has been reported that rats tested in the context of the usual pre-drug cues are more tolerant to the thermic (Siegel 1978*b*), locomotor (Mucha *et al.* 1981), and behaviourally-sedating (Hinson and Siegel 1983) effects of morphine than equally drug-experienced rats tested in the context of alternative cues.

Environmental specificity of tolerance to non-opiate drugs

Environmental specificity of tolerance is rather general, and has been demonstrated with a variety of non-opiate drugs. It has been demonstrated that Same-Tested mice (Melchior and Tabakoff 1981, 1985) and rats (Parker and Skorupski 1981) are more tolerant to the sedating effects of ethanol than Different-Tested subjects. Similar environmental specificity has also been reported with respect to tolerance to the hypothermic (Crowell *et al.* 1981; Lê *et al.* 1979; Mansfield and Cunningham 1980; Melchior 1986) and lethal (Melchior and Tabakoff 1982) effects of ethanol in animals, and a variety of effects of ethanol in humans (Annear and Vogel-Sprott 1985; Beirness and

Vogel-Sprott 1984; Dafters and Anderson 1982; Shapiro and Nathan 1986).

Environmental specificity of tolerance has been demonstrated with a variety of non-opiate drugs in addition to ethanol; amphetamine, haloperidol, and various benzodiazepines (Siegel 1986, 1987). Such findings have inspired an analysis of tolerance that emphasizes Pavlovian conditioning principles.

THE PAVLOVIAN CONDITIONING SITUATION

Living organisms respond not only reflexively to stimuli, they also respond in anticipation of stimuli. The analysis of such anticipatory responding uses procedures and terminology developed by Ivan Pavlov, and is called Pavlovian (or classical) conditioning (Pavlov 1927). In the Pavlovian conditioning situation, a contingency is arranged between two stimuli; typically, one stimulus reliably predicts the occurrence of the second stimulus. Using the usual terminology, the second of these paired stimuli is termed the 'unconditional stimulus' (UCS). The UCS, as the name implies, is selected because it elicits relevant activities from the outset (i.e., unconditionally), prior to any pairings. Responses elicited by the UCS are termed unconditional responses (UCRs). The stimulus signalling the presentation of the UCS is 'neutral' (i.e., it elicits little relevant activities prior to its pairing with the UCS), and is termed the 'conditional stimulus' (CS). The CS, as its name implies, becomes capable of eliciting new responses as a function of (i.e., conditional upon) its pairing with the UCS.

In Pavlov's well-known conditioning research, the CS was (for example) a bell, and the UCS was food (which elicited a conveniently monitored salivary response). Following a number of CS-UCS pairings, it was noted that the subject salivated not only in response to the UCS, but also in anticipation of the UCS (i.e., in response to the CS). The subject is then said to display a conditional response (CR).

Drugs as UCSs

A variety of interoceptive and exteroceptive stimuli have been used in Pavlovian conditioning experiments. Drugs constitute a particularly interesting class of UCSs. After some number of drug administrations, each administration reliably signalled by a CS, pharmacological CRs can be observed in response to the CS. It was Pavlov who first demonstrated such pharmacological conditioning; he paired a tone with administration of apomorphine. The drug induced restlessness, salivation, and a 'disposition to vomit'. After several tone-apomorphine pairings, the tone alone 'sufficed to produce all the active symptoms of the drug, only in a lesser degree ...' (Pavlov 1927, p. 35). Additional research by Krylov (reported by Pavlov 1927, p. 35–7) indicated that even if there is not an explicit CS (such as an auditory cue), naturally-occurring pre-drug cues (opening the box containing the hypodermic syringe, cropping the fur, etc.) could serve as CSs. In Krylov's experiments, a dog was repeatedly injected with morphine, each injection eliciting a number of responses including copious salivation. After five or six injections, it was observed that 'the preliminaries of injection' (Pavlov 1927, p. 35) elicited many morphine-like responses, including salivation.

Drug compensatory CRs

Most pharmacological conditioning research has been greatly influenced by Pavlov's theory of CR formation. According to this theory, the CR is a replica of the UCR, and, indeed, much drug conditioning work has demonstrated CRs that mimic the drug effect (Siegel 1985). In contrast, in 1937 Subkov and Zilov reported (p. 296) that dogs with a history of epinephrine administration (each injection eliciting a tachycardiac response), displayed a conditional bradycardiac response. Subkov and Zilov cautioned against 'the widely accepted view that the external modifications of the conditional reflex must always be identical with the response of the organism to the unconditional stimulus'.

Subsequent research and theory suggest that the characteristics of the pharmacological CR depend on the nature and mechanism of the unconditional drug effect (Eikelboom and Stewart 1982). For many effects of many drugs, the CR is an anticipatory compensation; the drug-associated environmental cues elicit responses that are opposite to the drug effect. For example, the subject with a history of morphine administration (and its analgesic consequence) displays a CR of hyper-algesia (Krank *et al.* 1981; Siegel 1975*b*). Similar compensatory-CRs have been reported with respect to the thermic (Siegel 1978*b*), locomotor (Mucha *et al.* 1981), behaviourally sedating (Hinson and Siegel 1983), and gastrointestinal (Raffa *et al.* 1982) effects of morphine. The CR seen with many non-opiate drugs is similarly opposite to the drug effect, for example, atropine (Mulinos and Lieb 1929), chlor-promazine (Pihl and Altman 1971), ethanol (Lê *et al.* 1979; Melchior 1986), methyl dopa (Korol and McLaughlin 1976), lithium chloride (Domjan and Gillan 1977), haloperidol (King *et al.* 1978), insulin (Siegel 1975*a*), and caffeine (Rozin *et al.* 1984).

COMPENSATORY CRs AND TOLERANCE

Drug-compensatory CRs are revealed by presentation of the usual pre-drug cues without the usual pharmacological consequences. When these cues are followed by the usual drug, the compensatory CR would be expected to attenuate the drug effect. As the association between the environmental CS and the pharmacological UCS is strengthened by repeated pairings, these CRs increasingly attenuate the drug effect. The progressively diminished response to a drug over the course of repeated administrations defines tolerance.

RELATIONSHIP BETWEEN CONDITIONING AND
NON-ASSOCIATIVE INTERPRETATIONS OF TOLERANCE

The conditioning analysis of tolerance is not an alternative to traditional interpretations. Rather, the conditioning analysis is complementary to views of tolerance that do not acknowledge a role for learning. Many such non-associative analyses of tolerance emphasize the role of drug-elicited homeostatic corrections that restore pharmacologically-induced physiological disturbances to normal levels. Several investigators have indicated the potential adaptive advantage if these homeostatic corrections actually antedate the pharmacological insult (e.g., Wikler 1973). Pavlov was certainly aware of the importance of such anticipatory responding; 'Under natural conditions the normal animal must respond not only to stimuli which themselves bring immediate benefit or harm, but also to other physical or chemical agencies—waves of light and the like—which in themselves only signal the approach of these stimuli' (Pavlov 1927, p. 14). On the basis of a conditioning model, the systemic alterations that mediate tolerance occur not only in response to the pharmacological stimulation, but may also occur in response to reliable environmental signals of this stimulation.

It should be apparent that conditioning is irrelevant to some prepara-tions that have been valuable in the study of tolerance. For example, tolerance studied in isolated tissue samples, or tolerance studied by continuous administration procedures (e.g., inhalation, or liquid diet) is not amenable to alteration by environmental manipulations (Melchior and Tabakoff 1984). On the other hand, conditioning processes are likely to contribute to tolerance under the usual conditions of drug administra-tion (regular administrations, at widely-spaced intervals, with discrete environmental cues uniquely present at the time of each administration), as it is these circumstances that favour the development of an association between pre-drug cues and the systemic effects of the drug.

EVIDENCE FOR CONDITIONING MODEL OF TOLERANCE

A considerable amount of evidence has been published that supports the conditioning analysis of tolerance. Much of these findings have been summarized elsewhere (Siegel 1983, 1985, in press *a*, in press *b*), thus the present review emphasizes more recent research, and only briefly summarizes earlier work to provide background for the more recent findings.

Environmental specificity of tolerance

The observation that there often is pronounced environmental specificity to the display of tolerance is readily interpretable by an analysis of tolerance that incorporates Pavlovian conditioning principles. If the repeatedly-drugged organism receives the drug in the context of the usual pre-drug cues, the compensatory CR partially cancels the drug effect, thus tolerance is observed. On the other hand, if this drug-experienced organism receives the drug in the context of cues not previously associated with the drug, there would be no pharmacological CR cancelling the drug effect, and the tolerance attributable to such a CR would not be observed.

Environmental specificity of tolerance and opiate overdose

The Conditioning Model of tolerance has been elaborated to account for some instances of overdose in human heroin addicts (Siegel 1984; Siegel and Ellsworth 1986; Siegel *et al.* 1982). Although deaths from overdoses are prevalent, the mechanisms of many of these deaths are far from clear. Some deaths result from the pharmacological overdoses (Huber 1974), but often victims die following doses that would not be expected to be fatal for these drug-experienced, and presumably drug-tolerant, individuals (Brecher 1972, p. 101–14; Reed 1980); indeed, the victims sometimes die following self-administration of a heroin dose that was well tolerated the previous day (Government of Canada 1973, p. 314). Some fatalities may result from a synergism between the opiate and other drugs concomitantly administered, or from adulterants (especially quinine) in the illicit heroin, but many deaths do not result from such drug interactions (Brecher 1972; Government of Canada 1973; Reed 1980). Thus, it has been suggested that 'the term "overdose" has served to indicate lack of understanding of the true mechanism of death in fatalities directly related to opiate abuse' (Greene *et al.* 1974, p. 175).

Some instances of these enigmatic failures of tolerance may be interpretable by the conditioning analysis. According to this analysis of overdose, an organism is at risk of overdose when the drug is administered in an environment that has not previously been extensively paired with the drug (and thus does not elicit the compensatory pharmacological CR that attenuates the effect of the drug).

Results of a recent experiment support the Pavlovian conditioning interpretation of heroin overdose (Siegel *et al.* 1982). Rats injected with high doses of heroin in the same environment as that previously associated with the drug were more likely to survive than rats with the identical pharmacological history receiving the final drug administration in an alternative environment.

The role of pre-drug cues in overdose has also been evaluated by interviewing drug addicts who have survivevd a heroin overdose. Findings obtained in such retrospective studies are mixed; Siegel (1984) has reported that novel pre-drug cues typically accompany such overdoses, but Neuman and Ellis (1986) reported that there is typically nothing unusual about the pre-drug cues on the occasion of the overdose.

A recent report suggests that Pavlovian conditioning may be relevant to some instances of death from overdose of licitly used opiates (Siegel and Ellsworth 1986). A single case is described, a patient receiving morphine for relief of pain from pancreatic cancer. The circumstances of this patient's death from apparent overdose of medically-prescribed morphine are readily interpretable by the Pavlovian conditioning account of tolerance.

Extinction of tolerance

Following CR acquisition, presentation of the CS without the UCS causes a decrease in response strength (i.e., 'extinction'). If drug tolerance is partially mediated by drug-compensatory CRs, extinction of these CRs should attenuate tolerance. That is, established tolerance should be reversed by placebo administrations. Such extinction has been demonstrated with respect to tolerance to the analgesic effect of morphine injected subcutaneously (e.g., Siegel *et al.* 1980) and directly into the ventricles of the brain (MacRae and Siegel, in press). Furthermore, extinction has been demonstrated with respect to tolerance to the lethal effect of morphine (Siegel *et al.* 1979), as well as a variety of effects of amphetamine, midazolam (a short-acting benzodiazepine), and the synthetic polynucleotide, Poly I:C (Dyck *et al.* 1986; Siegel 1986, 1987).

External inhibition of tolerance

Conditional responses, once established, can be disrupted by the presentation of a novel, extraneous stimulus. The phenomenon was termed 'external inhibition' by Pavlov (1927, p. 44) who described its operation in the salivary conditioning situation.

'The dog and the experimenter would be isolated in the experimental room, all the conditions remaining for a while constant. Suddenly some disturbing factor would arise—a sound would penetrate the room; some quick change in illumination would occur, the sun going behind a cloud; or a draught would get in underneath the door, and maybe bring some odour with it. If any of these extra stimuli happened to be introduced just at the time of application of the conditioned stimulus, it would inevitably bring about a more or less pronounced weakening or even a complete disappearance of the reflex response depending on the strength of the extra stimulus.'

On the basis of the conditioning analysis of tolerance, it would be expected that the display of tolerance should be disrupted merely by presentation of a novel stimulus. Results of a recent experiment (Siegel and Sdao-Jarvie 1986) supported this prediction with respect to tolerance to the hypothermic effect of ethanol. The rats used in this experiment were prepared with chronically implanted temperature telemetry devices; thus, the time course of drug-induced temperature alterations could be monitored without the necessity of subject handling. One group of rats was injected with ethanol (1 g/kg) for 12 consecutive sessions (the inter-session interval was 48 hr), and a second group was injected with physiological saline for 12 sessions. The thermic effect of these substances (depicted as temperature changes from pre-injection baselines) for the first and twelfth session are shown in the left and middle panels, respectively, of Figure 6.2. As is apparent in the left panel of Figure 6.2, the first injection of physiological saline was followed by a transitory increase in temperature (probably as a result of the stress induced by the injection procedure). The first injection of ethanol, however, had a pronounced hypothermic effect.

Tolerance to the hypothermic effect of ethanol was apparent by Drug Day 12; the drug no longer affected the temperature of the ethanol-experienced rats. As indicated in the middle panel, ethanol-injected and saline-injected subjects responded with almost the same pattern of post-injection temperature alteration.

The similar post-injection temperatures seen in the two groups on Drug Day 12, however, resulted from different processes. This was revealed by the effects of the photostimulus during the test session, Test Day 1, which was the session following Drug Day 12. Each group was again injected with its usual substance, ethanol or physiological saline, but, starting 20 minutes after injection, and continuing for 20 minutes, a novel stimulus was presented, a bright strobe light flashing at 4 Hz. The temperature changes seen following the injection of this test session are

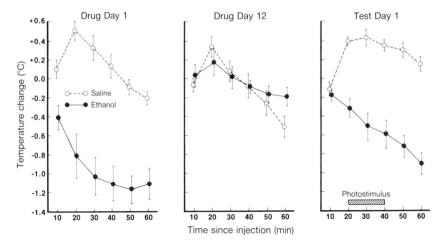

FIG. 6.2. Mean post-injection temperature changes (± 1 s.e.m.) following injection of saline or ethanol on Drug Day 1 (left panel), Drug Day 12 (middle panel), and the immediately subsequent Test Day (right panel). On the Test Day, a novel stimulus (provided by a photostimulator) was presented 20–40 minutes post-injection. (From Siegel and Sdao-Jarvie 1986.)

depicted in the right panel of Figure 6.2. The effect of the photostimulus was markedly different in the two groups; it exacerbated the hyperthermic effect of the saline injection, and reinstituted the hypothermic effect of the ethanol. Indeed, the effect of the extraneous stimulus persisted even following its termination; ethanol-injected subjects became progressively more hypothermic following the period of photostimulation application. This observation of the persistent after-effect of an external inhibiting stimulus confirms Pavlov's (1927, p. 45) observation that such a novel stimulus 'is effective not only while it lasts, but also for some time after its cessation while its after-effect lasts'.

These findings are consistent with the conditioning account of ethanol tolerance, that is, such tolerance is, at least in part, mediated by an association between environmental cues present at the time of drug administration and the systemic effects of the drug (although there are alternative interpretations, see Peris and Cunningham 1986).

Other evidence for the conditioning analysis of tolerance

In addition to the research summarized above, results of many other experiments have provided further evidence that Pavlovian conditioning contributes to tolerance to many drugs. These experiments demonstrate that nonpharmacological manipulations of pre-drug environmental cues affect both CR acquisition and tolerance in a similar manner. Tolerance to morphine and Poly I:C, in common with other CRs, can be retarded by extensive preconditioning experience with the CS (i.e., drug-predictive cues; Dyck *et al.* 1986; Siegel 1977; Tiffany and Baker 1981); tolerance to a variety of effects of morphine is retarded by partial reinforcement (i.e., pairing only a portion of the CS presentations with the UCS; Krank *et al.* 1984; Siegel 1977, 1978*b*). Tolerance to both morphine (Fanselow and German 1982; Siegel *et al.* 1981) and pentobarbital (Hinson and Siegel 1986) is subject to inhibitory learning. Morphine tolerance is subject to sensory preconditioning (Dafters *et al.* 1983), and can be manipulated by compound conditioning phenomena such as 'blocking' (Dafters *et al.* 1983) and 'overshadowing' (Dafters and Bach 1985; Walter and Riccio 1983). A full discussion of these findings is beyond the scope of this review, but it should be emphasized that a variety of additional experiments support the conditioning analysis of tolerance.

CUES FOR DRUGS

Injection cues

In most experiments involving the manipulation of pre-drug cues, these cues involve transporting the subject to a distinctive room at a time (typically 30 minutes to one hour) before the drug is administered in that room, or the presentation of distinctive audio-visual stimuli prior to each drug administration (Siegel 1983). In fact, in the absence of special precautions, the effects of a drug are almost always signalled by cues uniquely present at the time of drug administration. For example, piercing the skin with a hypodermic needle (in the case of parenterally administered drugs) reliably announces subsequent pharmacological stimulation, and there is evidence that this cue importantly contributes to tolerance to the analgesic effect of morphine (Dafters and Bach 1985).

Pharmacological cues

There is a considerable amount of evidence that drug states can serve as salient stimuli (Overton 1984). That is, responses learned when the subject is not under the influence of a centrally-acting drug may fail to be displayed when the subject is subsequently tested while under the influence of this drug. To the extent that tolerance is mediated by learning, it might be expected that tolerance will display such drug state-dependency.

Results of recent research concerning the effects of pentobarbitone on the display of morphine tolerance may be interpreted as resulting from the effects of a pharmacological signal (generated by the interoceptive effects of the barbiturate) and morphine. Terman and his colleagues (1983, 1985) reported that rats with a history of morphine administration, anaesthetized with pentobarbitone prior to a final injection of morphine, do not display the analgesic tolerance seen in non-anaesthetized rats; the barbiturate apparently blocks morphine tolerance. Recent experiments evaluated the relationship between the conditioning analysis of tolerance and the effect of pentobarbitone in blocking tolerance to the analgesic effect of morphine (Siegel, submitted). Specifically, the experiments were designed to determine whether the barbiturate affects morphine tolerance because it functions to alter cues signalling the opiate.

According to the state-dependency interpretation, the crucial feature of the observed effect of pentobarbitone on morphine tolerance is that the tolerance test involves an alteration in the effective cues from those prevailing during pre-test morphine administrations. This is to be contrasted with interpretations that hypothesize some direct pharmaco-dynamic interaction between the barbiturate and opiate (Pontani *et al.* 1985). On the basis of a state-dependency interpretation, it would be expected that subjects tested for morphine tolerance while under the influence of pentobarbitone, but also trained under that pharmacological state, should not show pentobarbitone-blockage of morphine tolerance (inasmuch as there is no change in the state signalling morphine). It would further be expected that the presence of pentobarbitone on the tolerance test should not be crucial for the blockage of tolerance. Rather, such blockage should also be seen in morphine-experienced subjects tested for analgesic tolerance without prior pentobarbitone if pre-test morphine administration were signalled by pentobarbitone (inasmuch as such a procedure involves alteration in the state signalling morphine). Both expectations were confirmed (Siegel, in press); rats that had pre-test administrations of morphine signalled by pentobarbitone, as well as the test administration, displayed substantial morphine tolerance. Moreover, if pre-test morphine administrations were signalled by pento-barbitone, omission of the barbiturate cue on the test sessions attenuated tolerance.

The pattern of results indicates that pentobarbitone affects morphine tolerance because it serves as a pre-drug signal. That is, just as there is environmental specificity of tolerance, there is also state specificity of tolerance (with the state being induced by pharmacologically-generated cues).

Early effect of a drug as a cue for later effect

As indicated above, the state induced by a drug can become a signal for a second, subsequently-administered drug. Another potential pharmacological signal for a drug is provided by the interoceptive effects of the drug itself. That is, the early effect of a drug inevitably signals a later effect. It has recently been demonstrated that the interoceptive effect of ethanol can serve as an effective cue for ethanol, and this association may contribute to tolerance (Greeley *et al.* 1984). In this study, rats in one group (Paired) consistently received a low dose of ethanol (0.8 g/kg) 60 minutes prior to a high dose (2.5 g/kg). Another group (Unpaired) received the low and high doses on an unpaired basis. When tested for tolerance to the hypothermic effect of 2.5 g/kg ethanol, Paired subjects, but not Unpaired subjects, displayed tolerance. Moreover, when Paired subjects received the high dose of ethanol not preceded by the low dose (on a test session), they failed to display their usual tolerance. This tolerance, dependent on ethanol-ethanol pairing, was apparently mediated by an ethanol-compensatory thermic CR; Paired rats, but not Unpaired rats, displayed a hyperthermic CR (opposite to the hypothermic effect of ethanol) in response to the low dose of ethanol. In addition, the tolerance seen in Paired rats was subject to extinction; repeated presentation of the low dose not followed by the high dose led to diminution of tolerance established in Paired rats.

Results of this Greeley *et al.* (1984) study provide convincing evidence that a small dose of a drug can serve as a signal for a larger dose of the same drug. Because a gradual increase in systemic concentration is an inevitable consequence of most drug administration procedures, such naturally-occurring drug-drug associations may play a hitherto unappreciated role in the effects of repeated drug administrations.

PAVLOVIAN CONDITIONING AND DRUG DEPENDENCE

According to most current views, tolerance and withdrawal symptoms are both manifestations of homeostatic mechanisms that correct for pharmacological disturbances; the feedback mechanisms that mediate tolerance (when the drug is administered) are expressed as withdrawal symptoms (when the drug is not administered). It has become increasingly apparent that just as drug anticipation contributes to tolerance, it also contributes to responses seen when drug administrations are terminated. That is, some drug 'withdrawal symptoms' are, more accurately, drug 'preparation symptoms'; they result from drug-compensatory CRs.

In discussing the role of compensatory CRs in so-called withdrawal symptoms, it is important to make a distinction between the acute withdrawal reaction seen shortly after the initiation of abstinence (which typically lasts for days or, at most, weeks) and the apparently similar symptoms often noted after detoxification is presumably complete (Hinson and Siegel 1982). In the latter case it is likely that it is the anticipation of the drug, rather than the drug itself, that is responsible for the symptoms.

'Consider the situation in which the addict expects a drug, but does not receive it; that is, no drug is available, but the addict is in an environment where he or she has frequently used drugs in the past, or it is the time of day when the drug is typically administered, or any of a variety of drug-associated stimuli occur. Research with animals demonstrates that presentation of cues previously associated with drug administration, but now not followed by the drug, results in the ocurrence of drug-compensatory CRs ... In the situation in which the drug addict expects but does not receive the drug, it would be expected that drug-compensatory CRs would also occur. These CRs normally counter the pharmacological disruption of functioning which occurs when the anticipated drug is administered. However, since the expected drug is not forthcoming, the CRs may achieve expression as overt physiological reactions, e.g., yawning, running nose, watery eyes, sweating ... or form the basis for the subjective experience of withdrawal sickness and craving' (Hinson and Siegel 1982, p. 499).

There is a substantial amount of clinical, experimental, and epidemiological evidence substantiating the contribution of drug-compensatory CRs to so-called withdrawal symptoms.

Clinical evidence

Many clinicians have commented on the ability of drug-associated environmental cues to elicit withdrawal symptoms in the long-detoxified former addict. For example, following a long period of enforced drug abstinence in prison, the former convict displays substantial withdrawal distress when he returns to his old neighbourhood, rich in drug-associated cues. Addicts asked to describe situations that elicit withdrawal-sickness typically describe circumstances closely associated with drug administration, for example, being offered some heroin by a friend. Indeed, a frequently-reported precursor to post-treatment relapse is confrontation with drug-associated stimuli (summarized in Siegel 1983).

Similar observations have been made in experimentally-addicted animals. For example, there is a description of the behaviour of monkeys that were repeatedly injected with morphine in the presence of tape-recorded music. This auditory stimulus became capable of eliciting withdrawal distress; 'After the animal had been weaned from the drug and maintained drug-free for several months, the experimenter again played the tape-recorded music and the animal showed the following signs; he became restless, had piloerection, yawned, became diuretic, showed rhinorrhea, and again sought out the drug injection' (Ternes 1977, pp. 167–8). As described more fully elsewhere (Siegel 1983), these symptoms seen in formerly opiate-dependent organisms, as well as 'withdrawal symptoms' in general, can be characterized as being opposite in direction to the drug effect.

Experimental evidence

Results of many experiments (reviewed in Siegel 1983) have confirmed the common clinical observation that drug-associated cues can elicit 'withdrawal symptoms'. Generally, these studies have demonstrated that, when addicts are confronted with drug-associated stimuli in the laboratory (e.g., the smell of alcohol for an alcoholic, or the injection paraphernalia for the intravenous heroin user), they display substantial symptoms of drug withdrawal (i.e. more than are displayed by non-clinical populations presented with these stimuli).

Some experiments with rats have similarly indicated that drug-associated environmental cues exacerbate 'withdrawal symptoms' (e.g., Hinson and Siegel 1983). Also, rats addicted to orally-consumed morphine, and then denied access to the drug, will subsequently relapse when the drug is again available in accordance with expectations of a conditioning analysis; that is, relapse is most pronounced when the environment of readdiction is most similar to the environment of original addiction (Hinson *et al.* 1986; Thompson and Ostlund 1965).

Epidemiological evidence

On the basis of a conditioning analysis, it would be expected that the prognosis for long-term success of addiction treatment would be greater if, following treatment, the individual returns to an environment other than that in which he or she originally became addicted. This alternative environment, unlike the addiction environment, would not contain a wealth of CSs for compensatory pharmacological CRs. In agreement with this prediction are the results of several studies indicating that residence relocation facilitates abstinence in opiate-dependent persons (Maddux and Desmond 1982).

Other epidemiological data consistent with the Conditioning Model have been obtained from analyses of relapse in United States Vietnam veterans, addicted to heroin while in Vietnam. These men were detoxified, discharged, and returned to the United States, to an environment very different than that in which they became addicted. Although a substantial social problem was envisaged (because of the high relapse rate typically seen in treated heroin addicts), relapse was not substantial in this group. These Vietnam veterans displayed a much lower relapse rate than did non-military heroin addicts, treated in federal facilities and returned to the same environment in which they were originally addicted (Robins *et al.* 1975).

There are, of course, many possible reasons why returning, heroin-addicted Vietnam veterans did not display substantial relapse (other than the conditioning rationale offered here). As is the case with correlational data, they are subject to a variety of interpretations. These epidemiological findings, however, are quite parallel with the experimental results concerning relapse in rats, and provide further evidence of the contribution of conditioning to drug dependence.

ROLE OF SELF-ADMINISTRATION

The Pavlovian conditioning analysis emphasizes the contribution of pre-drug cues to drug tolerance and dependence. In the discussion thus far, these cues have been conceptualized as environmental (i.e., the physical location of drug administration) or pharmacological (i.e., one drug signalling another, or the early effect of a drug signalling the later effect). Often, of course, drugs are self-administered. It might be expected that interoceptive cues accompanying self-administration (e.g., cognitive-volitional or proprioceptive signals for the systemic effect of the drug) similarly contribute to the effects of repeated pharmacological stimulation.

Self administration cues have been evaluated in experiments that compare the effects of drugs in animals that self-administer the drug with effects in animals yoked to these self-administering animals. Typically, the self-administering subject is prepared with a chronic jugular cannula, allowing for repeated intravenous injection. They can inject themselves with an opiate by pressing the lever in an experimental chamber. Yoked animals are similarly cannulated, and placed in a similar chamber, but lever presses have no consequence. Rather, the yoked subject receives the drug at the same time as the self-administering subject. Thus, yoked animals have no control over drug delivery, but rather receive the same doses of the drug, equally often, and at the same intervals as the self-administering animals.

There is evidence that the physiological effects of opiates are different in self-administering and yoked rats (Smith *et al.* 1982, 1984*a,b*). It has also recently been demonstrated that the severity of withdrawal symptoms is different in self-administering and yoked rats, despite their identical pharmacological histories (MacRae and Siegel, in preparation).

In this experiment demonstrating the different effects of self-administered and passively-received morphine (MacRae and Siegel, in preparation), rats participated in squads of three. For each experimental session, members of a simultaneously-run trio (each pre-experimentally implanted with a chronic intravenous cannula) were individually confined in one of three identical experimental chambers. Each subject's cannula was attached to an automatic syringe pump. For one member of a trio (assigned to the Self-Infused group), each operation of a lever in the chamber operated all three syringe pumps. The pumps delivered 0.69 mg of morphine sulphate to both the Self-Infused rat and a rat in another chamber (Yoked-Morphine). The third member of the trio (Yoked-Ringer's) received an infusion of an equivalent volume of the vehicle, Ringer's solution, whenever the Self-Infused and Yoked-Morphine rats received morphine infusions. The experiment was conducted in repeated seven-day cycles. For the first six days of each cycle, the drugs were administered in accordance with the above procedure. For the seventh day, no drugs were administered to any subject; rather, their withdrawal behaviour was systematically observed and scored in accordance with standard procedures (Gianutsos *et al.* 1975).

The various withdrawal behaviours observed in the three groups are depicted in Figure 6.3. As can be seen, there was much more pronounced withdrawal distress in Self-Infused subjects than in the equally morphine-experienced Yoked-Morphine subjects. It would appear that intero-

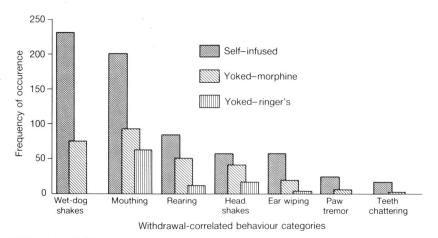

FIG. 6.3. Withdrawal behaviours in groups of rats that had self-administered morphine (Self-Infused), or had been yoked to these self-administrators and passively received either morphine (Yoked-Morphine) or Ringer's solution (Yoked-Ringers).

ceptive signals of a drug, incidental to voluntary self-administration, can importantly influence the magnitude of withdrawal symptoms.

CONCLUSIONS

It is well-established that drug tolerance is influenced by experience with drug-paired cues, as well as the drug. An interpretation of this influence emphasizes the interaction between learning and pharmacology; the organism learns, via Pavlovian conditioning, an association between pre-drug cues and the systemic effects of the drug. It has frequently been noted that conditional drug responses (elicited by presenting pre-drug signals without the drug) are opposite to the drug effect. According to the conditioning analysis, these conditional drug-compensatory responses contribute to tolerance by attenuating the drug effect when the drug is administered following the usual pre-drug cues. Results of many experiments support the conditioning analysis of tolerance by demonstrating that non-pharmacological manipulations of drug-predictive stimuli affect the acquisition of tolerance in much the same way as they affect the acquisition of Pavlovian conditional responses.

When the usual pre-drug signals are not followed by the usual pharmacological consequences, drug-compensatory conditional responses may sometimes be manifest as 'withdrawal symptoms'. There is a substantial amount of experimental, clinical, and epidemiological data that attest to the role of such conditional responses in so-called withdrawal symptoms.

The cues that become associated with drugs may be environmental (e.g., the location of drug administration), pharmacological (e.g., the early effect of a drug signalling a later effect), or internal states that accompany volitional self-administration. Thus, in addition to an understanding of pharmacodynamic and pharmacokinetic principles, a complete account of drug tolerance requires an appreciation of Pavlovian conditioning principles.

REFERENCES

Adams, W. J., Yeh, S. Y., Woods, L. A., and Mitchell, C. L. (1969). Drug-test interaction as a factor in the development of tolerance to the analgesic effect of morphine. *J. Pharmacol. and Exper. Ther.* **168**, 251–7.

Advokat, C. (1980). Evidence for conditioned tolerance of the tail flick reflex. *Behavioral and Neural Biology* **29**, 385–9.

Annear, W. C. and Vogel-Sprott, M. (1985). Mental rehearsal and classical conditioning contribute to ethanol tolerance in humans. *Psychopharmacology* **87**, 90–3.

Beirness, D. and Vogel-Sprott, M. (1984). Alcohol tolerance in social drinkers: Operant and classical conditioning effects. *Psychopharmacology* **84**, 393–7.

Brecher, E. M. (1972). *Licit and illicit drugs.* Little Brown, Boston.

Crowell, C. R., Hinson, R. E., and Siegel, S. (1981). The role of conditional drug responses in tolerance to the hypothermic effect of ethanol. *Psychopharmacology* **73**, 51–4.

Dafters, R. and Anderson, G. (1982). Conditioned tolerance to the tachycardia effect of ethanol in humans. *Psychopharmacology* **78**, 365–7.

—— and Bach, L. (1985). Absence of environment-specificity in morphine tolerance acquired in nondistinctive environments: Habituation or stimulus overshadowing? *Psychopharmacology* **87**, 101–6.

——, Hetherington, M., and McCartney, H. (1983). Blocking and sensory pre-conditioning effects in morphine analgesic tolerance: Support for a Pavlovian conditioning model of drug tolerance. *Quart. J. Exper. Psychology* **35B**, 1–11.

Domjan, M. and Gillan, D. J. (1977). After-effects of lithium-conditioned stimuli on consummatory behavior. *J. Exper. Psychology: Animal Behavior Processes* **3**, 322–34.

Dyck, D. G., Greenberg, A. H., and Osachuk, T. A. G. (1986). Tolerance to drug-induced (Poly I:C) Natural Killer (NK) cell activation: Congruence with a Pavlovian conditioning model. *J. Exper. Psychology: Animal Behavior Processes* **12**, 25–31.

Eikelboom, R. and Stewart, J. (1982). Conditioning of drug-induced physiological responses. *Psychological Review* **89**, 507–28.

Fanselow, M. S. and German, C. (1982). Explicitly unpaired delivery of morphine and the test situation: Extinction and retardation of tolerance to the suppressing effects of morphine on locomotor activity. *Behavioral and Neural Biology* **35**, 231–41.

Gianutsos, G., Drawbaugh, R., Hynes, M., and Lal, H. (1975). The narcotic withdrawal syndrome in the rat. In *Methods in narcotics research* (ed. S. Ehrenpreis and A. Neidle), pp. 293–309. Marcel Dekker, New York.

Government of Canada (1973). *Final report of the commission of inquiry into the nonmedical use of drugs.* Information Canada, Ottawa.

Greeley, J., Lê, D. A., Poulos, C. X., and Cappell, H. (1984). Alcohol is an effective cue in the conditional control of tolerance to alcohol. *Psychopharmacology* **83**, 159–62.

Greene, M. H., Luke, J. L., and Dupont, R. L. (1974). Opiate "overdose" deaths in the District of Columbia I. Heroin-related fatalities. *Med. Ann. of the District of Columbia* **43**, 75–181.

Hinson, R. E. and Siegel, S. (1982). Nonpharmacological bases of drug tolerance and dependence. *J. Psychosomatic Research* **26**, 495–503.

—— and —— (1983). Anticipatory hyperexcitability and tolerance to the narcotizing effect of morphine in the rat. *Behavioral Neuroscience* **97**, 759–67.

—— and —— (1986). Pavlovian inhibitory conditioning and tolerance to pentobarbital-induced hypothermia in rats. *J. Exper. Psychology: Animal Behavior Processes* **12**, 363–70.

——, Poulos, C. X., Thomas, W., and Cappell, H. (1986). Pavlovian conditioning and addictive behavior: Relapse to oral self-administration of morphine. *Behavioral Neuroscience* **100**, 368–75.

Huber, D. H. (1974). Heroin deaths—Mystery or overdose? *J. Amer. Med. Assoc.* **229**, 689–90.

Kavaliers, M. and Hirst, M. (1986). Environmental specificity of tolerance to morphine-induced analgesia in a terrestrial snail: Generalization of a behavioral model of tolerance. *Pharmacology Biochemistry and Behavior* **25**, 1201–6.

King, J. J., Schiff, S. R., and Bridger, W. H. (1978). Haloperidol classical conditioning—Paradoxical results. *Soc. for Neuroscience Abstracts* **4**, 495.

Korol, B. and McLaughlin, L. J. (1976). A homeostatic adaptive response to alpha-methyl-dopa in conscious dogs. *Pavlovian J. Biolog. Science* **11**, 67–75.

Krank, M. D., Hinson, R. E., and Siegel, S. (1981). Conditional hyperalgesia is elicited by environmental signals of morphine. *Behavioral and Neural Biology* **32**, 148–57.

——, ——, and —— (1984). The effect of partial reinforcement on tolerance to morphine-induced analgesia and weight loss in the rat. *Behavioral Neuroscience* **98**, 79–85.

Lê, A. D., Poulos, C. X., and Cappell, H. (1979). Conditioned tolerance to the hypothermic effect of ethyl alcohol. *Science* **206**, 1109–10.

MacRae, J. and Siegel, S. (in press). Extinction of tolerance to the analgesic effect of morphine: Intracerebroventricular administration and effects of stress. *Behavioral Neuroscience*.

—— and —— (in preparation). Differential effects of morphine in self-administering and yoked-control rats.

Maddux, J. F. and Desmond, D. P. (1982). Residence relocation inhibits opioid dependence. *Archives of General Psychiatry* **39**, 1313–17.

Mansfield, J. G. and Cunningham, C. L. (1980). Conditioning and extinction of tolerance to the hypothermic effect of ethanol in rats. *J. Comparative and Physiol. Psychology* **94**, 962–9.

Melchior, C. L. (1986). Environment-dependent tolerance generated with intracerebroventricular injections of ethanol. *Alcoholism: Clinical and Experimental Research* **10**, 116.

—— and Tabakoff, B. (1981). Modification of environmentally-cued tolerance to ethanol in mice. *J. Pharmacol. and Exper. Ther.* **219**, 175–80.

—— and —— (1982). Environment-dependent tolerance to the lethal effects of ethanol. *Alcoholism: Clinical and Experimental Research* **6**, 306.

—— and —— (1984). A conditioning model of alcohol tolerance. In *Recent developments in alcoholism*, Vol. 2 (ed. M. Galanter), pp. 5–16. Plenum Press, New York.

—— and —— (1985). Features of environment-dependent tolerance to ethanol. *Psychopharmacology* **87**, 94–100.

Mucha, R. F., Volkovsiks, C., and Kalant, H. (1981). Conditioned increases in locomotor activity produced with morphine as an unconditioned stimulus, and the relation of conditioning to acute morphine effect and tolerance. *J. Comparative and Physiol. Psychology* **95**, 351–62.

Mulinos, M. G. and Lieb, C. C. (1929). Pharmacology of learning. *Amer. J. Physiology* **90**, 456–7.

Neumann, J. K. and Ellis, A. R. (1986). Some contradictory data concerning a behavioral conceptualization of drug overdose. *Bull. of the Soc. of Psycho-*

logists in Addictive Behavior **5**, 87–90.

Overton, D. A. (1984). State dependent learning and drug discriminations. In *Handbook of psychopharmacology*, Vol. 18 (eds. L. L. Iversen, S. D. Iversen, and S. H. Snyder), pp. 59–127. Plenum Press, New York.

Parker, L. F. and Skorupski, J. D. (1981). Conditioned ethanol tolerance. In *Drug dependence and alcoholism* (ed. A. J. Schecter), pp. 665–71. Plenum Press, New York.

Pavlov, I. P. (1927). *Conditioned reflexes* (trans. G. V. Anrep). Oxford University Press, London.

Peris, J. and Cunningham, C. L. (1986). Handling-induced enhancement of alcohol's acute physiological effects. *Life Sciences* **38**, 273–9.

Pihl, R. O. and Altman, J. (1971). An experimental analysis of the placebo effect. *J. Clin. Pharmacol.* **11**, 91–5.

Pontani, R. B., Vadlamani, N. L., and Misra, A. L. (1985). Potentiation of morphine analgesia by subanesthetic doses of pentobarbital. *Pharmacol. Biochemistry and Behavior* **22**, 395–8.

Raffa, R. B., Porreca, F., Cowan, A., and Tallarida, R. J. (1982). Evidence for the role of conditioning in the development of tolerance to morphine-induced inhibition of gastrointestinal motility in rats. *Federation Proceedings* **41**, 1317.

Reed, T. (1980). Challenging some "common wisdom" on drug abuse. *Internat. J. Addictions* **15**, 359–73.

Robins, L. N., Helzer, J. E., and Davis, D. H. (1975). Narcotic use in southeast Asia and afterwards. *Arch. General Psychiatry* **32**, 955–61.

Rozin, P., Reff, D., Mark, M., and Schull, J. (1984). Conditioned responses in human tolerance to caffeine. *Bull. Psychonomic Society* **22**, 117–20.

Shapiro, A. P. and Nathan, P. E. (1986). Human tolerance to alcohol: The role of Pavlovian conditioning processes. *Psychopharmacology* **88**, 90–5.

Siegel, S. (1975*a*). Conditioning insulin effects. *J. Comparative and Physiol. Psychology* **89**, 189–99.

—— (1975*b*). Evidence from rats that morphine tolerance is a learned response. *J. Comparative and Physiol. Psychology* **89**, 498–506.

—— (1976). Morphine analgesic tolerance: Its situation specificity supports a Pavlovian conditioning model. *Science* **193**, 323–5.

—— (1977). Morphine tolerance acquisition as an associative process. *J. Exper. Psychology: Animal Behavior Processes* **3**, 1–13.

—— (1978*a*). Is there evidence for a conditioning model? *Science* **200**, 344–5.

—— (1978*b*). Tolerance to the hyperthermic effect of morphine in the rat is a learned response. *J. Comparative and Physiol. Psychology* **92**, 1137–49.

—— (1983). Classical conditioning, drug tolerance, and drug dependence. In *Research advances in alcohol and drug problems*, Vol. 7, (eds. Y. Israel, F. B. Glaser, H. Kalant, R. E. Popham, W. Schmidt, and R. G. Smart), pp. 207–46. Plenum Press, New York.

—— (1984). Pavlovian conditioning and heroin overdose: Reports by overdose victims. *Bull. Psychonomic Society* **22**, 428–30.

—— (1985). Drug-anticipatory responses in animals. In *Placebo: Theory, research and mechanisms* (eds. L. White, B. Tursky, and G. E. Schwartz), pp. 288–305. Guilford Press, New York.

—— (1986). Environmental modulation of tolerance: Evidence from benzo-diazepine research. In *Tolerance to beneficial and adverse effects of antiepileptic drugs* (eds. H. Meinardi, W. Fröcher, H. H. Frey, and W. P. Koella), pp. 89–100. Raven Press, New York.

—— (1987). Pavlovian conditioning and ethanol tolerance. *Alcohol and Alcoholism*, Supplement 1, 25–36.

—— (in press). State dependent learning and morphine tolerance. *Behavioral Neuroscience.*

—— and Ellsworth, D. (1986). Pavlovian conditioning and death from apparent overdose of medically prescribed morphine: A case report. *Bull. Psychonomic Society* **24**, 278–80.

—— and MacRae, J. (1984). Environmental specificity of tolerance. *Trends in Neuroscience* **7**, 140–2.

—— and Sdao-Jarvie, K. (1986). Reversal of ethanol tolerance by a novel stimulus. *Psychopharmacology* **88**, 258–61.

——, Hinson, R. E., and Krank, M. D. (1978). The role of predrug signals in morphine analgesic tolerance: Support for a Pavlovian conditioning model of tolerance. *J. Exper. Psychology: Animal Behavior Processes* **4**, 188–96.

——, ——, and —— (1979). Modulation of tolerance to the lethal effect of morphine by extinction. *Behavioral and Neural Biology* **25**, 257–62.

——, Sherman, J. E., and Mitchell, D. (1980). Extinction of morphine analgesic tolerance. *Learning and Motivation* **11**, 289–301.

——, Hinson, R. E., and Krank, M. D. (1981). Morphine-induced attenuation of morphine tolerance. *Science* **212**, 1533–4.

——, ——, ——, and McCully, J. (1982). Heroin "overdose" death: The contribution of drug-associated environmental cues. *Science* **216**, 436–7.

Smith, J. E., Co, C., Freeman, M. E., and Lane, J. D. (1982). Brain neurotransmitter turnover correlated with morphine-seeking behavior of rats. *Pharmacol. Biochemistry and Behavior* **16**, 509–19.

——, ——, and Lane, J. D. (1984*a*). Limbic acetylcholine turnover rates correlated with rat morphine-seeking behaviors. *Pharmacol. Biochemistry and Behavior* **20**, 429–42.

——, ——, and —— (1984*b*). Limbic muscarinic cholinergic and benzodiazepine receptor changes with chronic morphine self-administration. *Pharmacol. Biochemistry and Behavior* **20**, 443–50.

Subkov, A. A. and Zilov, G. N. (1937). The role of conditioned reflex adaptation in the origin of hyperergic reactions. *Bull. Biologie et de Médecine Experimentale* **4**, 294–6.

Terman, G. W., Lewis, J. W., and Liebskind, J. C. (1983). Sodium pentobarbital blocks morphine tolerance and potentiation in the rat. *Physiologist* **26**, A–111.

——, Pechnick, R. N., and Liebskind, J. C. (1985). Blockade of tolerance to morphine analgesia by pentobarbital. *Proc. Western Pharmacol. Soc.* **28**, 157–60.

Ternes, J. W. (1977). An opponent process theory of habitual behavior with special reference to smoking. In *National Institute on Drug Abuse Research Monograph* 17, pp. 157–82. United States Government Printing Office, Washington DC.

Thompson, T. and Ostlund, W. (1965). Susceptibility to readdiction as a function of the addiction and withdrawal environments. *J. Comparative and Physiological Psychology* **60**, 388–92.

Tiffany, S. T. and Baker, T. B. (1981). Morphine tolerance in the rat: Congruence with a Pavlovian paradigm. *J. Comparative and Physiol. Psychology* **95**, 747–62.

Walter, T. A. and Riccio, D. C. (1983). Overshadowing effects in the stimulus control of morphine analgesic tolerance. *Behavioral Neuroscience* **97**, 658–62.

Wikler, A. (1973). Conditioning of successive adaptive responses to the initial effects of drugs. *Conditional Reflexes* **8**, 193–210.

7

Drug reward studied by the use of place conditioning in rats

CHRISTINA SPYRAKI

INTRODUCTION

The place conditioning (PC) procedure is used to evaluate the motivational properties of drugs. It was introduced early this decade to compensate for methodological and interpretive difficulties associated with the self-administration technique, the conventional method for assessing reinforcing properties of drugs (Capell and Le Blanc 1981; Schuster and Johanson 1981). Using this paradigm, the primary reinforcing properties of a drug are conditioned to salient environmental stimuli which, by association, acquire secondary reinforcing properties. Saline injections are paired with different stimuli and, on the test day, the animal, in the undrugged state, is allowed to 'choose' between the stimuli. This technique only requires that the animals perform a simple operation to approach or avoid the place paired with the drug. If the drug experience is one of positive affect, it is presumed that subjects will spend more time in the place of the drug experience.

After half a decade of studies with the place conditioning procedure its use appears to be debatable, although, because of its simplicity, the method has become very popular. The aim of this presentation is to examine if the initial aspiration of using place conditioning as an 'ideal' tool to study drug reward has been satisfied.

VARIABLES AFFECTING DRUG-INDUCED PLACE CONDITIONING (PC)

Procedural variables

Biased or unbiased procedure
The biased procedure refers to the use of a pre-conditioning test to estimate initial preferences of animals. The PC score is expressed as the

difference between pre- and post-conditioning time spent on the drug-associated side of the training box. In the unbiased procedure, the side of conditioning is determined by random assignment of animals to one of the two compartments counterbalanced across treatment groups. No differences have been noticed in the PC score by researchers who used both procedures for a case (Blander *et al.* 1984; Mucha and Iversen 1984; Spyraki *et al.* 1985; Nomikos and Spyraki 1987). However, in some cases those procedures may yield discrepant results. The biased procedure seems to be the more powerful method, as it makes it more difficult for the animal to show a preference for the drug-conditioned side of the box (initially least preferred) and thus bias the paradigm against the display of a strong preference.

Conditioning on the non-preferred or preferred side

When the biased procedure is used, animals are routinely conditioned on the non-preferred side for preference and on the preferred side for aversion, depending on the prediction which has been made for the substance under study. The motivational properties of the substance have to be shown regardless of where the conditioning took place. However, this is not always the case. It appears that heroin conditioning in the preferred environment is not effective (Schenk *et al.* 1985), and this holds true for cocaine given intraperitoneally, but not intravenously (Nomikos and Spyraki 1987). Although theoretical and methodological interpretation of those peculiar results may exist, more detailed studies are needed to examine the issue.

Amount of training during conditioning

The amount of training given to each rat depends on the conditioning time of each trial, and on the number of trials. The time of each conditioning trial does not appear to be a critical factor (Mucha *et al.* 1982) unless it is extended over the effective time period of a drug (Sherman *et al.* 1980*b*). The number of training trials, depending on the route of administration, is a determinant factor influencing the magnitude of place conditioning score (Blander *et al.* 1984). Thus, one single trial is required for animals to show place preference following intravenous (IV) morphine (Mucha *et al.* 1982) or IV cocaine (Nomikos and Spyraki 1987), two trials for subcutaneous (SC) cocaine (Isaak *et al.* 1984), three to four trials for SC morphine or naloxone (Mucha and Iversen 1984) and intraperitoneal (IP) cocaine (Nomikos and Spyraki 1987). However, Bardo and his colleagues (1986) obtained place preference (PP) following one single intra-peritoneal (IP) injection of cocaine. Their animals were not pre-exposed to the test apparatus (unbiased procedure). Considering that IV administration yields faster and stronger effect than the IP administration, the results show that place conditioning is a method sensitive to the amount of reward.

Behavioural variables

The preference or aversion for the drug-associated side of the training box is the result of a discrimination between drug and vehicle state. Thus, during conditioning, place discrimination learning is taking place. The strength of the conditioned response (CR), preference or aversion, is not measured directly, but it is inferred from an operant choice of behaviour which is brought under stimulus control, and it is reflected in the increased or decreased duration that an animal spends in the presence of drug-associated stimuli. It is then apparent that behavioural variables may influence the expression of PC behaviour and interfere with the interpretation of the results.

Temporal contiguity of drug–environment associations

If the drug is given after the animal has experienced the to-be-conditioned environment, place conditioning does not occur. This is evident in studies with morphine (Bardo *et al.* 1984; Mucha and Iversen 1984; Sherman *et al.* 1980*a*), amphetamine (Sherman *et al.* 1980*b*), cocaine (Nomikos and Spyraki 1987), and diazepam (Spyraki *et al.* 1985). It is apparent that temporal overlap between the to-be-conditioned environment (conditioned stimulus) and the drug experience (unconditioned stimulus) is required.

Acquisition, extinction, and retention of the conditioned response (CR)

It has been made clear that the acquisition of the place conditioned response (PCR) is proportional to the amount of conditioning. In a study with cocaine (Bardo *et al.* 1986), in which an unbiased procedure was used, place conditioning tests intermittent between conditioning trials (partial reinforcement effect) have resulted in an attenuation of the strength of conditioning. Many studies using the biased procedure failed to observe such an effect (Blander *et al.* 1984; Phillips and LePiane 1980; Spyraki *et al.* 1987; Stapleton *et al.* 1979). The extinction of the CR to morphine, following six 30 min daily extinction-training sessions, was studied by Bardo and his colleagues (1986). They reported decreased per-entry duration and increased number of entries to the drug-associated side. This latter observation is reminiscent of the transient increase in lever pressing by rats often seen during extinction of drug self-administration. The CR to morphine or naloxone is retained up to thirty days (Mucha and Iversen 1984). Similarly, cocaine place preference has been observed four, seven, and 30 days following conditioning (Nomikos and Spyraki 1987).

State-dependent learning and novelty

During training the rats experience the drug-paired cues under the influence of the drug, while during the post-conditioning test, animals are in a drug-free state. In this state, the drug-paired environmental stimuli were not experienced during conditioning and the animals could be approaching them because of stimulus novelty. Similarly, because the drug-free state is probably reminiscent of a vehicle day, they could be avoiding them, approaching the vehicle-associated compartment (state-dependent learning).

To check for state dependency, place conditioning is also assessed under the drug condition. To assess novelty-induced motivation on the post-conditioning test day, groups of unbiased animals experience only one compartment of the training box during conditioning. Studies with morphine (Mucha and Iversen 1984), diazepam (Spyraki *et al.* 1985), and cocaine (Nomikos and Spyraki 1987) have ruled out state-dependency and novelty as alternative explanations of drug-place conditioning.

Memory

It is obvious, and it has been elegantly shown (White and Carr 1985), that memory-improving or memory-disrupting properties of unconditioned stimulus (drugs under study) may influence PC. Thus a substance with amnesic properties, or injections into brain areas involved in memory processes, may give false negative results.

Conditioned behavioural effects

An alternative interpretation of place conditioning suggests that place-preference behaviour reflects the acquisition of a behavioural conditioned response (CR) rather than an appetitive CR. Such an interpretation is supported by the observation that environmental stimuli associated with opiates (Mansfield *et al.* 1981; Mucha *et al.* 1981; Sherman 1979; Wilder and O'Brien 1980) or psychostimulants (Barr *et al.* 1983; Beninger and Hahn 1983; Tilson and Rech 1973) can elicit conditioned responses which mimic the unconditioned effects of the drug.

Swerdlow and Koob (1984), showing that only freely-moving, unrestricted animals display a preference for the amphetamine-associated environment, have postulated that the motor-activating properties of drugs contribute to their place-conditioning properties. In favour of this hypothesis are findings showing that drugs increasing locomotion, such as amphetamine (Reicher and Holman 1977; Sherman *et al.* 1980*b*; Spyraki *et al.* 1982*c*), cocaine (Spyraki *et al.* 1982*b*),

D-Ala^2-Met^5-enkephalinamide (Phillips and LePiane 1982; Phillips *et al.* 1982), heroin (Bozarth and Wise 1981; Spyraki *et al.* 1983), morphine (Mucha *et al.* 1982; van der Kooy *et al.* 1982; Mucha and Iversen 1984), neurotensin (Glimcher *et al.* 1984), and substance P (Holzhäuer-Oitzl, *et al.* 1987) induce place preferences. Drugs like naloxone or vasopressin induce both spontaneous hypoactivity and place aversions (Ettenberg *et al.* 1983; Mucha and Iversen 1984; Pert and Sivit 1977). However, this interpretation is not supported by recent reports showing that drugs in doses which either do not influence activity, like methylphenidate (Martin-Iverson *et al.* 1985; Mithani *et al.* 1986), or induce hypoactivity, like morphine (Bardo *et al.* 1984), diazepam (File 1986; Spyraki *et al.* 1985) or clonidine (Asin and Wirtshafer 1985) also induce place preferences. Furthermore, treatments which abolish hyperactivity are ineffective in influencing place preferences (Di Scala *et al.* 1985; Martin-Iverson *et al.* 1985; Spyraki *et al.* 1982*b*), and treatments which induce hyperactivity also decrease place preferences (Nomikos *et al.* 1986).

These data suggest that a dissociation can be made between locomotor and rewarding processes. However, it must be considered that the dissociation may be based on data collected from experiments in which the 'non-specific' behavioural effects (locomotion) were measured following place-conditioning tests and outside the PC setting. Such an approach is not sufficient. Conditioned responses to drugs show environmental specificity and, by definition, they are not elicited in the absence of conditioned stimulus (in a novel environment). Thus, a modification of the place-conditioning apparatus, which would allow simultaneous assessment of behavioural effects of drugs (locomotion, rearings, stereotypy, analgesia, catalepsy, temperature) during conditioning and on the test day, is to be recommended. The data obtained from such an approach will probably not offer the answer to the main question, but at least screening for behavioural effects will be obtained as additional indexes of drugs' reinforcing potential.

Pharmacological variables

Route of administration

As is expected, the pharmacokinetic properties of a drug will influence both the appearance and the magnitude of the place-conditioning behaviour. Thus, naloxone is not effective when given by the IP route (Mucha and Iversen 1984; Mucha *et al.* 1982; Phillips and LePiane 1980; Spyraki *et al.* 1985), and it appears that IV injections of drugs yield greater place preference scores than IP injections (Mucha *et al.* 1982; Mucha and Iversen 1984; Nomikos and Spyraki 1987).

Time between drug injection and conditioning

In many studies, the time between drug injection and onset of conditioning, chosen arbitrarily, corresponds to the time by which a known behavioural effect of the substance under study is seen.

The issue has been addressed directly only in two studies, one with morphine (Rossi and Reid 1976) and the other with alcohol (Reid *et al.* 1985). Emphasizing the importance that such a variable may have on place conditioning, both studies point to the need for further research in order to determine the optimum time lapse between injection and onset of conditioning.

Dose–response curves

Recent detailed studies, in contrast to the initial ones (Mucha *et al.* 1982; Spyraki *et al.* 1982*b,c*), have shown dose-related place-conditioning behaviours with a variety of opiate receptor agonists and antagonists (Barr *et al.* 1985; Mucha and Iversen 1984; Mucha and Herz 1985; Amalric *et al.* 1987) and with IV cocaine (Nomikos and Spyraki 1987).

With the majority of substances, single phase dose–response curves are obtained. However, a U-shaped dose–response curve is obtained with IV cocaine (Nomikos and Spyraki 1987) and with the kappa opiate receptor agonist U50-488 (Mucha and Herz 1985). The biphasic function of dose, showing a reversal of the effect with higher doses, may be attributable to additional activity of the substance on a different substrate. As drugs of abuse possess both rewarding and punishing properties, the U-shaped dose response curve appears of great interest. It points once more to the sensitivity of the method of assessing appetitive and aversive properties of a substance, and to the need for further studies employing a very wide dose range of the substance under study.

Antecedent conditions

Place preference (PP) with ethanol can only be shown with animals previously exposed to the drug (Reid *et al.* 1985). In contrast, in the case of morphine, PP is not shown if animals are pre-exposed to both drug and experimental environment (Mucha *et al.* 1982).

Furthermore, rats isolated at weaning appear less sensitive to heroin or cocaine in the place-conditioning procedure (Schenk *et al.* 1985, 1986).

Organismic variables

There are anecdotal and scattered reports of individual variations in the place-conditioning performance of animals in experimental and control groups (Reid *et al.* 1985; Schenk *et al.* 1986). This represents a weakness of place conditioning; it can be circumvented by running controls concurrently with experimental groups, and by including large number of subjects per group.

At this moment, it is unknown if differences in the strain, species, age, and sex of the animals may influence place conditioning.

PLACE CONDITIONING: A METHOD TO SCREEN DRUGS FOR ABUSE LIABILITY

Drugs inducing place preference

Table 7.1 shows that a variety of drugs of different drug classes induce place preference (PP). Those drugs are abused by humans, and are self-administered by rodents and non-human primates. Assuming that PP is a conditioned response to appetitive properties of drugs, it is apparent that the procedure is valid for the assessment of positive reinforcement, and thus useful for screening drugs for abuse liability. It should be mentioned that reward induced by natural reinforcers, such as food to hungry animals, can also be detected by the method (Holzäuer and Huston, personal communication; Spyraki *et al.* 1982*a*).

The advantage of the method over self-administration is stressed mainly by the observation that diazepam induces place preference (File 1986; Spyraki *et al.* 1985). Diazepam is abused by humans, but rodents have difficulties initiating and maintaining self-administration of the drug. Probably the rewarding properties of the drug are masked, in an operant method, by its muscle-relaxant actions.

It is apparent that, in contrast to the rate-dependent, motorially-demanding operant procedures, the place-conditioning method is not sensitive to the 'non-specific' drug effects which may be of primary importance in modulating response rate performances.

A variety of drugs, listed on Table 7.2, induce place aversion. This points to the capability of the method to reveal aversive properties of certain substances. This is another reason for making the PC method of study advantageous over the self-administration method. The latter cannot differentiate between motivationally neutral and aversive effects of drugs. The animal eventually ceases pressing the lever, regardless of whether the stimulus is neutral or aversive.

In addition, the conditioning shows stereospecificity, receptor specificity and dose specificity. Only the active isomers, (−)morphine and levorphanol, induce place preference (Mucha and Herz 1986), while the active isomer (−)bremazocine induces place aversion. Mu opiate receptor agonists induce place preference, while kappa agonists induce

TABLE 7.1. *Drug-induced place preference*

A Opiates		
	Morphine	Advokat 1985; Barr *et al.* 1985; Blander *et al.* 1984; Katz and Gormezano 1979; Mackey and van der Kooy 1985; Mucha and Iversen 1984; Mucha *et al.* 1982; Rossi and Reid 1976; Sherman *et al.* 1980*a*; Stapleton *et al.* 1979
	(−)Morphine	Mucha and Herz 1986
	Levorphanol	Mucha and Herz 1986
	Heroin	Bozarth and Wise 1981; Schenk *et al.* 1985; Spyraki *et al.* 1983
	Etorphine	Mucha *et al.* 1982
	Fentanyl	Mucha and Herz 1985
	Sufentanyl	Mucha and Herz 1985
	Methylnatrexone	Bechera and van der Kooy 1985
B CNS stimulants		
	d-Amphetamine	Gilbert and Cooper 1983; Mackey and van der Kooy 1985; Reicher and Holman 1977; Schenk *et al.* 1986; Sherman *et al.* 1980*b*; Spyraki *et al.* 1982*c*; Swerdlow and Koob 1984; White and Carr 1985
	l-Amphetamine	Gilbert and Cooper 1983
	β-phenylethylamine	Gilbert and Cooper 1983
	Cocaine	Bardo *et al.* 1986; Mackey and van der Kooy 1985; Isaak *et al.* 1984; Schenk *et al.* 1986; Spyraki *et al.* 1982*b*)
	Methylphenidate	Martin-Iverson *et al.* 1985; Mithani *et al.* 1986
	Nomifensine	Martin-Iverson *et al.* 1985
	Bupropion	Ortmann 1985
	Apomorphine	Spyraki *et al.* 1982*c*; van der Kooy *et al.* 1983
	Nicotine	Fudala *et al.* 1985
C Anxiolytics		
	Ethanol	Reid *et al.* 1985
	Diazepam	File 1986; Spyraki *et al.* 1985
	Lorazepam	File 1986
	Alprazolam	File 1986
	Adinazolam	File 1986
	Meprobamate	Spyraki *et al.* 1985
	Tracazolate	File 1986
	U-43,465	File 1986

place aversion (Mucha and Herz 1985). Methylnaltrexone or morphine, depending on the dose, produce place preference or place aversions (Bechera and van der Kooy 1985).

The data in Table 7.3 show a variety of central acting drugs which are

TABLE 7.2. *Drug-induced place aversion*

Lithium chloride	Mucha and Herz 1985; Mucha *et al.* 1982
Naloxone	Mucha and Iversen 1984; Mucha and Herz 1985; Mucha *et al.* 1982; Spyraki *et al.* 1985; van der Kooy *et al.* 1982
Naltrexon	Bechera and van der Kooy 1985
Morphine	Bechera and van der Kooy 1985
(−)Bremazocine	Mucha and Herz 1985
U 50-488	Mucha and Herz 1985
Pentobarbital	Mucha and Iversen 1984
Ethanol	Cunningham 1981
CGS 8216	File 1986
Picrotoxin	File 1986; Spyraki *et al.* 1985
Yohimbine	File 1986
Scopolamine	MacMahon *et al.* 1981

TABLE 7.3. *Drugs not inducing place conditioning*

Haloperidol	Martin-Iverson *et al.* 1985; Spyraki *et al.* 1982*a,b,c*
Pimozide	Bozarth and Wise 1981; Spyraki *et al.* 1982*b*
Flupenthixol	Mackey and van der Kooy 1985
Domperidone	Spyraki and Fibiger 1987
Desmethylimipramine	Martin-Iverson *et al.* 1985
Zimeldine	Kruszewska *et al.* 1986
Progabide	Di Scala *et al.* 1985
Sodium Valproate	Spyraki *et al.* 1985
CGS 8216	File 1986; Spyraki *et al.* 1985
Ethanol	Asin *et al.* 1985; Stewart and Grupp 1986; van der Kooy *et al.* 1983
Chlordiazepoxide	File 1986
Buspirone	File 1986
Naloxone	Bozarth and Wise 1981; Mucha and Iversen 1984; Phillips and LePiane 1980

not effective in place conditioning, at least for the doses tested. It is well known that all drugs of abuse are discriminated. Moreover, many psychoactive drugs (neuroleptics, antidepressants) not abused by humans, also induce stimulus change discriminable by man and by animals. This observation weakens the validity of operant drug discrimination procedure as a method to screen drugs for abuse liability. In view of the data in Table 7.3, the place-conditioning method, which also involves learning discrimination, has the advantage over operant drug discrimination procedures of being less sensitive to the non-rewarding effects of the drugs.

The usefulness of a method, especially when it has to be employed in drug screening, is also dictated by its cost. For the self-administration method, only a few animals are required to screen reinforcing properties of drugs. Each animal can (and must) serve as its own control. In order to study a substance for place conditioning properly, hundreds of animals are required, and hundreds of hours! It is necessary either to watch the animal during the experiment, or a videotape of its responses. Thus, although the method for studying place conditioning is valid to assess motivational properties of drugs, it is not 'cost and time efficient'.

PLACE CONDITIONING: A METHOD TO STUDY NEURAL SUBSTRATES OF DRUG REWARD

Data on drug reinforcement, gained through the self administration technique, are of great value. However, the results are confounded by the fact that, in operant tasks, subjects are required to perform a complex motor response (often signalled by a tone or light) while under the influence of a pharmacological or neurological treatment that may impair motor performance or perception. There are procedures for circumventing some of these self-injection problems, but the fact remains that these findings merit verification with other paradigms.

The conditioned place preference (PP) paradigm is ideally suited to the analysis of the neural mechanisms underlying drug reinforcement because the reward effect is indicated by a preference test after the drug has worn off. A neuropharmacological analysis of brain systems involved in drug reinforcement simply requires pre-treatment with a specific antagonist prior to each attempt to establish conditioning reinforcement to environmental stimuli paired with the rewarding drug. Subjects treated with saline should show a significant preference for the conditioning stimuli when tested subsequently in an undrugged state. In contrast, subjects in which the primary reinforcing effects of the drug were blocked by pre-treatment with the antagonist should not show this preference conditioning. Other approaches used to delineate the neural substrate of drug reward have employed intracerebral injections of drugs in specific brain areas, or chemically-induced lesions of different neuronal systems.

Psychostimulants

Haloperidol pre-treatment antagonizes the place preference (PP) produced by d-amphetamine (Mackey and van der Kooy 1985; Spyraki *et al.* 1982*c*). In addition, in rats that had received 6-OHDA-induced lesions of the nucleus accumbens (NAS), the time spent in the amphetamine-associated environment correlates significantly with the level of dopamine (DA) remaining in the NAS but not with the DA contained in the striatum (Spyraki *et al.* 1982*c*). Carr and White (1983) reported that amphetamine injections into NAS but not into the striatum produces place preference. These findings support the view that the rewarding effects of d-amphetamine are mediated by central dopamine neurons and, in particular, by those of the mesolimbic system. However, the same was not so for cocaine (Mackey and van der Kooy 1985; Spyraki *et al.* 1982*b*), methylphenidate and nomifensine (Martin-Iverson *et al.* 1985), or bupropion (Ortman 1985). All those drugs are stimulants but, unlike amphetamine, are dopamine-uptake inhibitors. Dopaminergic mechanisms appear to be involved in the self administration of cocaine (Roberts *et al.* 1980) and nomifensine (Spyraki and Fibiger 1981).

In addition, 6-OHDA-induced destruction of central or peripheral noradrenergic systems did not affect cocaine (Spyraki *et al.* 1982*b*) or amphetamine (Spyraki *et al.* 1982*c*) place preference.

To overcome the problem of local anaesthetic action that would interfere with cocaine place preference (Spyraki *et al.* 1982*b*), and to minimize the differences in experimental conditions between place-preference and self-administration methods, animals were used, in the place-conditioning procedure, with permanent catheters in the right jugular vein through which they received cocaine. It was observed that IV cocaine-induced place preference was totally blocked by haloperidol (Nomikos and Spyraki 1987). It is of interest that ICV cocaine place preference is blocked by spiroperidol (Morency and Beninger 1987). This evidence delineates the role of dopamine in cocaine reinforcement; it remains for the specific dopaminergic system involved to be elucidated.

Opiates

The place conditioning paradigm provides a variety of data in support of an enkephalinergic substrate of reward. Place conditioning has been obtained with ICV injections of D-Ala^2-Met^5-enkephalin (Katz and Gormezano 1979), and there have been reports of place conditioning following localized ICV injections of morphine into the lateral hypothalamus, the nucleus accumbens (NAS), and periaqueductal gray, as well as into the ventral tegmental area (VTA) (Phillips and LePiane 1980; van der Kooy *et al.* 1982). Intrathecal administration of morphine induces place preference, involving spinal opiates in the reinforcing properties of morphine (Advokat 1985).

Place conditioning induced by unilateral injection of D-Ala^2-Met^5-enkephalin into the VTA was attenuated by systemic injections of the dopamine receptor blocker, haloperidol (Phillips *et al.* 1982). In addition, selective lesions of the ascending dopamine pathways ipsilateral to the injection site blocked the rewarding effect when dopamine levels were reduced by more than 90 per cent. Similar lesions in the contralateral hemisphere had no influence on this behaviour (Phillips *et al.* 1983). Furthermore, place preference (PP) induced by systemic morphine or heroin is attenuated by neuroleptics (Bozarth and Wise 1981; Spyraki *et al.* 1983; Schwartz and Marchok 1974). 6-OHDA lesions of the nucleus accumbens also attenuated heroin place preference (Spyraki *et al.* 1983). In view of the reversal of morphine PP by naloxone (Mucha and Iversen 1984; Mucha *et al.* 1982) and the blockade of intra-tegmental (VTA) morphine induced PP by naloxone, the data suggest that enkephalinergic-dopaminergic mechanisms, especially in the vicinity of the VTA, mediate opioid-induced reward. However, the hypothesis is opposed by the observation that morphine-induced PP is not influenced by flupenthixol pre-treatment (Mackey and van der Kooy 1985). In favour of the dopamine hypothesis of reward, it was found recently that injections of neurotensin, a non-opioid peptide, into the VTA induces place preference (Glimcher *et al.* 1984).

Diazepam

Pre-treatment with CGS 8216 antagonizes the effect of diazepam involving central benzodiazepine receptors in the appetitive properties of the drug (Spyraki *et al.* 1985). Pre-treatment with haloperidol antagonizes the diazepam place preference (PP). This is assumed to be a central antidopaminergic effect, as pre-treatment with domperidone (which does not cross the blood brain barrier) failed to influence this effect of diazepam. In addition, animals with 6-OHDA lesions in the nucleus accumbens do not show a preference for the compartment of the apparatus associated with diazepam (Spyraki and Fibiger, 1987). These findings suggest that dopamine containing neurons of the mesolimbic system are a component of the neural circuitry that mediates the appetitive properties of diazepam.

Beyond the mesolimbic dopaminergic system

The data presented so far point to a critical role of the dopaminergic innervation of the nucleus accumbens (NAS) in the positive reinforcing properties of amphetamine, heroin, and diazepam as these are expressed in the place-preference paradigm.

However, other neurotransmitters may also play a role in the functional output of the NAS. Gamma-aminobutyric acid (GABA) would be a putative candidate, but pre-treatment with progabide, a GABA-mimetic drug, failed to influence amphetamine place preference (Di Scala *et al.* 1983). Increased apomorphine-induced place preference following 6-OHDA lesions of the NAS (van der Kooy *et al.* 1983) disappeared following electrolytic lesios of substantia innominata (Swerdlow *et al.* 1984), an area to which NAS heavily projects.

To gain more insight into the role of NAS in drug reward, studies were made of amphetamine and morphine place preference in animals which suffered extensive depletions of 5-HT into the NAS following injections of the selective neurotoxin, 5,7 DHT. A significant attenuation of morphine-induced (but not amphetamine-induced) place preference was observed (Nomikos *et al.* 1986). Amphetamine place preference was slightly increased in animals treated either with p-chloroamphetamine (5-HT depletion) or with the 5-HT_2 antagonist, ritanserin (Nomikos and Spyraki, in preparation). Morphine-induced place preference was attenuated following similar treatments (Nomikos and Spyraki, in preparation). Pre-treatment with zimelidine, which increases serotonin transmission in the brain, blocks the place preference induced by amphetamine, leaving morphine place preference unaffected (Kruszewska *et al.* 1986). These data taken together suggest that serotonin differentially influences morphine and amphetamine place preference. They are also consistent with results from self-administration studies showing increased self-administration of amphetamine by the 5-HT antagonist metergoline and following intraventricular but not intra-accumbens injections of 5,7-DHT (Lyness and Moore 1983; Lyness *et al.* 1980).

CONCLUSION

Place conditioning procedures have been used to assess appetitive and aversive properties of drugs, and to study brain substrates mediating drug reward in the rat. A number of drugs of different classes have been shown to induce place conditioning.

Available data on the variables affecting drug-induced place conditioning are:

(1) procedural, including the use of animals pre-exposed or not to the test apparatus, conditioning on the most or on the least preferred environment, and duration of conditioning training;

(2) behavioural, such as temporal UCS-CS associations, acquisition, retention, extinction of the conditioned response, state-dependent learning, stimulus novelty effect, and concurrent conditioned behaviours;

(3) pharmacological, such as the route of administration, the time lapse

between drug injection and onset of conditioning, dose, and pharmacological history.

Many studies have been made using the place-conditioning procedure as a tool to delineate the neurochemical substrate mediating drug reward. Place conditioning induced by stimulants, opiates, and diazepam appears to be attenuated following blockade of dopamine receptors or 6-OHDA lesions of the nucleus accumbens. In contrast, drug-induced place conditioning is not affected by 6-OHDA-induced noradrenaline depletion. The available data support the view that dopaminergic neurons, mainly those of the mesolimbic system, may be involved in drug reward.

It is apparent, from the data presented above, that the use of place conditioning to study neural substrates of drug reward has yielded results confirming the conclusions reached through the self-administration method. This points to the validity of the method.

It is our hope that the method will also prove suitable in the analysis of neurochemical mechanisms underlying aversive properties of drugs. The method will be of great help in elucidating whether reward and aversion have a common or different substrate.

REFERENCES

Advokat, C. (1985). Evidence of place conditioning after chronic intrathecal morphine in rats. *Pharmacology, Biochemistry and Behavior* **22**, 271–7.
Amalric, M., Cline, E. J., Martinez, J. L. Jr., Bloom, F. E., and Koob, G. F. (1987). Rewarding properties of β-endorphin as measured by conditioned place preference. *Psychopharmacology* **91**, 14–19.
Asin, K. E. and Wirtshafter, D. (1985). Clonidine produces a conditioned place preference in rats. *Ppsychopharmacology* **85**, 383–5.
——, ——, and Tabakoff, B. (1985). Failure to establish a conditioned place preference with ethanol in rats. *Pharmacology, Biochemistry and Behavior* **22**, 169–73.
Bardo, M. T., Miller, J. S., and Neisewander, J. L. (1984). Conditioned place preference with morphine: The effect of extinction training on the reinforcing CR. *Pharmacology, Biochemistry and Behavior* **21**, 545–9.
——, Neisewander, J. L., and Miller, J. S. (1986). Repeated testing attenuates conditioned place preference with cocaine. *Psychopharmacology* **89**, 239–43.
Barr, G. A., Sharpless, N. S., Cooper, S., Schiff, S. R., Paredes, W., and Bridger, W. H. (1983). Classical conditioning, decay and extinction of cocaine-induced hyperactivity and stereotypy. *Life Sciences* **33**, 1341–51.
——, Paredes, W., and Bridger, W. H. (1985). Place conditioning with morphine and phencyclidine: dose dependent effects. *Life Sciences* **36**, 363–8.
Bechera, A. and van der Kooy, D. (1985). Opposite motivational effects of endogenous opioids in brain and periphery. *Nature* **314**, 533–4.
Beninger, R. and Hahn, B. L. (1983). Pimozide blocks establishment but not expression of amphetamine-produced environment specific conditioning.

Science **220**, 1304–6.

Blander, A., Hunt, T., Blair, R., and Amit, Z. (1984). Conditioned place preference: An evaluation of morphine's positive reinforcing properties. *Psychopharmacology* **84**, 124–7.

Bozarth, M. A. and Wise, R. A. (1981). Heroin reward is dependent on a dopaminergic substrate. *Life Sciences* **29**, 1881–6.

Capell, H. and Le Blanc, A. E. (1981). Tolerance and physical dependence: Do they play a role in alcohol and drug self administration? In *Research advances in alcohol and drug problems*, 6 (eds. Y. Israel, F. B. Glaser, H. Kalant, R. E. Popham, W. Schmidt, and R. G. Smart). Plenum Press, New York.

Carr, G. D. and White, N. M. (1983). Conditioned place preference from intra-accumbens but not intra-caudate amphetamine injections. *Life Sciences* **33**, 2551–7.

Cunningham, C. L. (1981). Spatial aversion conditioning with ethanol. *Pharmacology, Biochemistry and Behavior* **14**, 263–4.

Di Scala, G., Martin-Iverson, M. T., Phillips, A. G., and Fibiger, H. C. (1985). The effect of progabide (SL 76002) on locomotor activity and conditioned place preference induced by d-amphetamine. *European J. Pharmacol.* **107**, 271–4.

Ettenberg, A., van der Kooy, D., Le Moal, M., Koob, G. F., and Bloom, F. E. (1983). Can aversive properties of peripherally-injected vasopressin account for its putative role in memory? *Brain Research* **7**, 331–50.

File, S. E. (1986). Aversive and appetitive properties of anxiogenic and anxiolytic agents. *Behavioural Brain Research* **21**, 189–94.

Fudala, P. J., Teoh, K. W., and Iwamoto, E. T. (1985). Pharmacological characterization of nicotine-induced conditioned place preference. *Pharmacology, Biochemistry and Behavior* **22**, 237–41.

Gilbert, D. and Cooper, S. J. (1983). β-phenylethylamine, d-amphetamine and l-amphetamine-induced place preference conditioning in rats. *European J. Pharmacol.* **95**, 311–14.

Glimcher, P. W., Margolin, D. H., Giovino, A. A., and Hoebel, B. G. (1984). Neurotensin: a New "Reward Peptide". *Brain Research* **291**, 119–24.

Holzhäuer-Oitzl, M.-S., Boucke, K., and Huston, J. P. (1987). Reinforcing properties of substance P in the lateral hypothalamus revealed by conditioned place preference. *Pharmacol. Biochem. Behav.* (in press).

Isaak, W. L., Neisewander, J. L., Landers, T., Alkala, R. M., and Bardo, M. T. (1984). Mesocortical dopamine system lesions disrupt cocaine reinforced conditioned place preference. *Soc. for Neurosciences Abstracts* **10**, 126.

Katz, R. J. and Gormezano, G. (1979). A rapid and inexpensive technique for assessing the reinforcing effects of opiate drugs. *Pharmacology, Biochemistry and Behavior* **11**, 231–3.

Kruszewska, A., Romandini, S., and Samanin, R. (1986). Different effects of zimelidine on the reinforcing properties of d-amphetamine and morphine on conditioned place preference in rats. *European J. Pharmacol.* **125**, 283–6.

Lyness, W. H. and Moore, K. E. (1983). Increased self administration of d-amphetamine in rats pretreated with metergoline. *Pharmacology, Biochemistry and Behavior* **18**, 721–4.

——, Friedle, N. M., and Moore, K. E. (1980). Increased self administration of d-amphetamine after destruction of 5-hydroxy-tryptaminergic nerves. *Pharmacology, Biochemistry and Behavior* **12**, 937–41.

Mackey, W. B. and van der Kooy, D. (1985). Neuroleptics block the positive reinforcing effects of amphetamine but not of morphine as measured by place conditioning. *Pharmacology, Biochemistry and Behavior* **22**, 101–5.

MacMahon, S. W., Blampied, N. M., and Hughes, R. N. (1981). Aversive stimulus properties of scopolamine. *Pharmacology, Biochemistry and Behavior* **15**, 389–92.

Mansfield, J. G., Wenger, J. R., Benedict, R. S., Halter, J. B., and Woods, S. C., (1981). Sensitization to the hyperthermic and catecholamine-releasing effects of morphine. *Life Sciences* **29**, 1697–704.

Martin-Iverson, M. T., Ortmann, R., and Fibiger, H. C. (1985). Place preference conditioning with methylphenydate and nomifensine. *Brain Research* **332**, 59–67.

Mithani, S., Martin-Iverson, M. T., Phillips, A. G., and Fibiger, H. C. (1986). The effects of haloperidol on amphetamine- and methylphenidate-induced conditioned place preferences and locomotor activity. *Psychopharmacology* **90**, 247–52.

Morency, M. A. and Beninger, R. J. (1986). Dopaminergic substrates of cocaine-induced place conditioning. *Brain Res.* **399**, 33–41.

Mucha, R. F. and Herz, A. (1985). Motivational properties of kappa and mu opioid receptor agonists studied with place and taste preference conditioning. *Psychopharmacology* **86**, 274–80.

—— and —— (1986). Preference conditioning produced by opioid active and inactive isomers of levorphanol and morphine in rat. *Life Sciences* **38**, 244–9.

—— and Iversen, S. D. (1984). Reinforcing properties of morphine and naloxone revealed by conditioned place preference: a procedural examination. *Psychopharmacology* **82**, 241–7.

——, Volkovskis, C. and Kalant, G. (1981). Conditioned increases in locomotor activity produced with morphine as an unconditioned stimulus and the relation of the conditioning to acute morphine effect and tolerance. *J. Compar. and Physiol. Psychol.* **95**, 351–62.

——, van der Kooy, D., O'Shaughnessy, M., and Bucenieks, P. (1982). Drug reinforcement studied by the use of place conditioning in rat. *Brain Research* **243**, 91–105.

——, Millan, M. J., and Herz, A. (1985). Aversive properties of naloxone in non-dependent (naive) rats may involve blockade of central β-endorphin. *Psychopharmacology* **86**, 281–5.

Nomikos, G. G. and Spyraki, C. (1987). Cocaine-induced place conditioning: importance of route of administration and other procedural variables. *Psychopharmacology* (in press).

—— and ——. The effect of the 5-HT$_2$ antagonist ritansenin on drug-induced place preference. *Pharmacol. Biochem. Behav.* (submitted).

——, ——, Galanopoulou, P., and Papadopoulou, Z. (1986). Amphetamine and morphine induced place preference in rats with 5,7-dihydroxytryptamine lesions of the nucleus accumbens. *Psychopharmacology* **89**, S80.

Ortmann, R. (1985). The conditioned place preference paradigm in rats: effect of bupropion. *Life Sciences* **37**, 2021–7.

Pert, A. and Sivit, C. (1977). Neuroanatomical focus for morphine and enkephalin-induced hypermotility. *Nature* **265**, 645–7.

Phillips, A. G. and LePiane, F. G. (1980). Reinforcing effects of morphine microinjection into the ventral tegmental area. *Pharmacology, Biochemistry and Behavior* **12**, 965–8.

—— and —— (1982). Reward produced by micro-injection of (D-Ala²),Met⁵-enkephalinamide into the ventral tegmental area. *Behavioural Brain Research* **5**, 225–9.

——, Spyraki, C., and Fibiger, H. C. (1982). Conditioned place preference with amphetamine and opiates as reward stimuli: Attenuation by haloperidol. In *The neural basis of feeding and reward* (eds. B. G. Hoebel and D. Novin) pp. 455–64. Haer Institute, Brunswick, Maine.

——, LePiane, F. G., and Fibiger, H. C. (1983). Dopaminergic mediation of reward produced by direct injection of enkephalin into the ventral tegmental area of the rat. *Life Sciences* **33**, 2505–11.

Reicher, M. A. and Holman, E. W. (1977). Location preference and flavor aversion reinforced by amphetamine in rats. *Animal Learning and Behavior* **5**, 343–6.

Reid, L. D., Hunter, G. A., Beaman, C. M., and Hubbell, C. L. (1985). Toward understanding ethanol's capacity to be reinforcing: a conditioned place preference following injections of ethanol. *Pharmacology, Biochemistry and Behavior* **22**, 483–5.

Roberts, D. C. S., Koob, G. F., Klonoff, P., and Fibiger, H. C. (1980). Extinction and recovery of cocaine self administration following 6-hydroxydopamine lesions of the nucleus accumbens. *Pharmacology, Biochemistry and Behavior* **12**, 781–7.

Rossi, N. A. and Reid, L. D. (1976). Affective states associated with morphine injections. *Physiolog. Psychology* **4**, 269–74.

Schenk, K. S., Ellison, F., Hunt, T., and Amit, Z. (1985). An examination of heroin conditioning in preferred and non-preferred environments and in differentially housed mature and immature rats. *Pharmacology, Biochemistry and Behavior* **22**, 215–20.

——, Hunt, T., Malovechko, R., Robertson, A., Klukowski, G., and Amit, Z. (1986). Differential effects of isolation housing on the conditioned place preference produced by cocaine and amphetamine. *Pharmacology, Biochemistry and Behavior* **24**, 1793–6.

Schuster, C. R. and Johanson, C. E. (1981). An analysis of drug seeking behavior in animals. *Neuroscience and Biobehavioral Reviews* **5**, 315–23.

Schwartz, A. S. and Marchok, P. L. (1974). Depression of morphine seeking behaviour by dopamine inhibition. *Nature* **248**, 257–8.

Sherman, J. E. (1979). The effects of conditioning and novelty on the rats analgesic and pyretic responses to morphine. *Learning and Motivation* **10**, 383–418.

Sherman, J. E., Pickman, C., Rice, A., Liebeskind, J. C., and Holman, E. W. (1980a). Rewarding and adversive effects of morphine: Temporal and pharmacological properties. *Pharmacology, Biochemistry and Behavior* **13**, 501–5.

——, Roberts, T., Roskam, S. E., and Holman, E. W. (1980b). Temporal properties of the rewarding and aversive effects of amphetamine in rats. *Pharmacology, Biochemistry and Behavior* **13**, 597–9.

Spyraki, C. and Fibiger, H. C. (1981). Intravenous self administration of nomifensine in rats: Implications for abuse potential in humans. *Science* **212**,

1167–58.

—— and ——. A role of the mesolimbic dopamine system in the reinforcing properties of diazepam. *Psychopharmacology* (in press).

——, —— and Phillips, A. G. (1982*a*). Attenuation by haloperidol of place preference conditioning using food reinforcement. *Psychopharmacology* **77**, 379–82.

——, ——, —— (1982*b*). Cocaine induced place preference conditioning: Lack of effects of neuroleptics and 6-hydroxydopamine lesions. *Brain Research* **253**, 195–203.

——, ——, —— (1982*c*). Dopaminergic substrates of amphetamine-induced place preference conditioning. *Brain Research* **253**, 185–93.

——, ——, —— (1983). Attenuation of heroin reward in rats by disruption of the mesolimbic dopamine system. *Psychopharmacology* **79**, 278–83.

——, Kazandjian, A. and Varonos, D. D. (1985). Diazepam induced place preference conditioning: Appetitive and antiaversive properties. *Psychopharmacology* **87**, 225–32.

——, Nomikos, G. G., and Varonos, D. D. (1987). Intravenous cocaine-induced place preference: Attenuation by haloperidol. *Behav. Brain Res.* (in press).

Stapleton, J. M., Lind, M. D., Merriman, V. J., Bozarth, M. A., and Reid, L. D. (1979). Affective consequences and subsequent effects on morphine self administration of D-ala²-methionine enkephalin. *Physiolog. Psychol.* **7**, 146–52.

Stewart, R. B. and Grupp, L. A. (1986). Conditioned place aversion mediated by orally self-adminstered ethanol in the rat. *Pharmacology, Biochemistry and Behavior* **24**, 1369–75.

Swerdlow, N. R. and Koob, G. F. (1984). Restrained rats learn amphetamine-conditioned locomotion but not place preference. *Psychopharmacology* **84**, 163–6.

——, Swanson, L. W., and Koob, G. F. (1984). Electrolytic lesions of the substantia innominata and lateral preoptic area attenuates the "supersensitive" locomotor response to apomorphine resulting from denervation of the nucleus accumbens. *Brain Research* **306**, 141–8.

Tilson, H. A. and Rech, R. H. (1973). Conditioned drug effects and absence of tolerance to d-amphetamine induced motor activity. *Pharmacology, Biochemistry and Behavior* **1**, 149–53.

van der Kooy, D., Mucha, R. F., O'Shaughnessy, M., and Bucenicks, A. (1982). Reinforcing effects of brain microinjections of morphine revealed by conditioned place preference. *Brain Research* **243**, 107–17.

——, Swerdlow, N. R., and Koob, G. F. (1983). Paradoxical reinforcing properties of apomorphine: Effects of nucleus accumbens and area postrema lesions. *Brain Research* **259**, 111–18.

Vezina, P. and Stewart, J. (1984). Conditioning and place-specific sensitization of increases in activity induced by morphine in the VTA. *Pharmacology, Biochemistry and Behavior* **20**, 925–34.

Wilder, M. B. and O'Brien, J. H. (1980). Evoked potential conditioning using morphine as unconditioned stimulus in rats. *Exper. Neurology* **67**, 534–53.

White, N. M. and Carr, G. D. (1985). The conditioned place preference is affected by two independent reinforcement processes. *Pharmacology, Biochemistry and Behavior* **23**, 37–42.

8

The role of opiate mechanisms in social relationships

DAVID BENTON

INTRODUCTION

At all stages of human life the establishing and maintaining of intimate interpersonal attachments are critically important for healthy psychological functioning. Our evolutionary background, particularly that associated with the mammalian form of reproduction, has bred into us the need for social bonding. It is thus reasonable to expect that the forming of social attachments reflects neural circuits with a distinctive neurochemistry. The importance of a warm relationship between the human infant and its parent is increasingly recognized as being important for healthy psychological development. The absence or breaking of such a secure parent/child relationship has been associated with a range of disorders including depression, suicide, and an over-dependent personality. It is surprising that, given its obvious importance, the neurochemical bases of social attachments have been ignored to a large extent. The first objective of this chapter is to argue that the study of the neurochemical modulation of social relationships may give us an insight into the aetiology of various psychiatric disorders. The second aim is to report a series of experiments where the role of the various opiate receptors in different social relationships have been examined.

AN EVOLUTIONARY PERSPECTIVE

Human evolution has been associated with the ability to make tools and weapons, the development of verbal and written communication, abstract thinking, co-operation in hunting, and flexible social organization. One fundamental factor, without which all these achievements would not have been possible, is the development of the mammalian system of reproduction. Mammals specialize in quality rather than quantity, the so called K reproductive strategy; they produce small numbers of young after a long gestation, and expend significant amounts of time and energy in protecting and nourishing them.

When the life patterns of primates are examined, a very striking feature is the very large proportion of the life that is spent in infancy and adolescence; from monkey, to non-human primate, to the human animal the proportion of life spent in pre-adult stages increases progressively. It is during this long developmental stage that social, cultural, linguistic, and manual skills are acquired. If the mammalian strategy of feeding the young from the mother's body was to be successful, then other adaptions were necessary; for example, the tendencies for the mother to be attracted to the young, and for the young to bond with the parents, must have developed.

During our development as hunters and gatherers, if the mother's movements were to be limited by a long pregnancy and period of lactation, then bonding between males and females can be viewed as essential to the safe development of the offspring (Passingham 1982). Although plant materials were important in the diet of our ancestors, the development of the ability to kill game and hunt dangerous prey in organized packs created an important selection pressure. As humans are physically poorly-equipped to attack large prey, such activities required the development, during childhood, of skills (such as the use of weapons), social organization, co-operation, and the use of the intellectual skills that are facilitated by language. The acknowledgement of a social hierarchy, possibly developed from earlier mechanisms associated with dominance and submission, would have been necessary for organized hunting.

ATTACHMENT AND PSYCHIATRIC HEALTH

All psychoanalytic work stresses the importance of parent–child relationships in psychological development. Particularly after the seminal work of Bowlby (1965), the importance of a lasting, warm relationship between infant and parent has been a widely acknowledged factor in the development of a psychologically healthy human adult. In making this generalization, it should be recognized that the many methodological problems in this area have led to controversy. However, the better-designed studies and critical reviews (Bowlby 1980) support the general proposition that the breaking of a bond during childhood predisposes to later psychiatric problems. Parental loss during childhood has been associated with depression (Brown 1982), suicide (Adam 1982), and over-dependency (Birtchnell 1975) in the adult. Bowlby (1980) concluded that 'those who have suffered a childhood bereavement are not only rather more prone than others to develop a psychiatric disorder but that both the form and the severity of any disorder they develop is likely to be strongly influenced . . .'.

Henderson (1982) found that adults who viewed their social relationships as inadequate had a substantially increased risk of developing neurotic symptoms. A cycle of abuse has been described in which neglect during childhood leads to an adult who in turn abuses their own young (Fontana 1973). More positively, Argyle (1986) reported that a satisfying social relationship was one of the most frequently mentioned causes of happiness in the adult human. Although the disruption of bonds undoubtedly plays a role in many conditions, it is fair to say that we remain uncertain as to the magnitude of its influence, and the conditions under which it has an effect. In the case of depression, for example, theories increasingly emphasize mutli-dimensional causality, and perhaps we should not expect a simple, or too large, relationship between any one factor and a disorder.

The systemic examination of the physiological changes that occur when social bonds are broken may well tell us something about the nature of psychiatric disorders. This approach was foreseen by Darwin (1872) who suggested that adult grief had much in common with a child's response when separated from an adult, and with the forms of grief displayed in other species. Hofer (1984) similarly pointed out that the symptoms associated with bereavement in the adult resemble those that follow a separation of the infant from the mother. The behavioural changes include social withdrawal, sadness, sleep disturbance, and decreased food intake.

The neurochemistry of attachment

To date, much of the work concerning the neurochemistry of social behaviour has examined the role of endogenous brain opioids (Panksepp *et al.* 1978). Brief periods of social isolation increase pain sensitivity and decrease the analgesic efficacy of morphine (Panksepp 1980), whereas longer periods of isolation increase the reaction to morphine (Kostowski *et al.* 1977). Some of these effects have been interpreted as the consequences of opioid-receptor proliferation or supersensitivity, reflecting the absence of a socially-stimulated release of endogenous opioids. Petkov and his colleagues (1985) found that social isolation changed the number of opiate receptors in various areas of the brain. Alleva and his co-workers (1986) found that the sex of the companions of mice between birth and weaning influenced the reaction to morphine. A prediction from the hypothesis that opiates satisfy social needs is that the administration of these drugs should cause animals to spend their time further away from a conspecific, a prediction confirmed in both rats (Panksepp *et al.* 1979) and guinea pigs (Herman and Panksepp 1978). A comment concerning almost all the work in this field to date is that it has failed to reflect the complexity of the opioid system; there have been few attempts to distinguish the roles of the various opioid receptors.

The hormonal factors associated with rodent maternal behaviour have attracted some attention. However, with the exception of opioid mechanisms, other neurochemical factors have been little examined. Grimm and Bridges (1983) speculated that the action of progesterone and oestradiol may be mediated via increases in hypothalamic beta-endorphin. They found that morphine prevented the onset of maternal behaviour in the rat.

Very low doses of morphine reduce separation calling in puppies, chickens (Panksepp *et al.* 1978), and guinea pigs (Herman and Panksepp 1978). Centrally-administered alpha, beta, and gamma endorphin also reduce the calling of chicks. Beta endorphin is the only opioid that is consistently more potent in this respect than morphine (Panksepp *et al.* 1978). Naloxone has been found to increase the calling of young mice (Robinson *et al.* 1985), although in chicks this occurs only during the last third of the light and last two thirds of the dark phase of the daily cycle (Panksepp *et al.* 1978). Panksepp suggested that separation vocalization is particularly sensitive to opiates, although Hard and Engel (1986) found that the ultrasonic calling of mice pups was modulated by serotonergic mechanisms. Although similar reports are not available using rodents, in the chick Rossi and colleagues (1983) found the alpha 2 adrenoceptor agonist, clonidine, suppressed separation calling. These few preliminary studies of the pharmacological manipulation of separation vocalization are very promising; there are clear suggestions of its mediation by particular transmitter systems. There remains the need to examine drugs that act via the sub-types of opiate, dopamine, noradrenaline, and 5-HT receptors, systematically.

Autism, an endorphin-mediated disorder?

Both Kalat (1978) and Panksepp (1979) have pointed out that there are many similarities between the influence of opiates and autism, leading them to suggest that the disorder may be caused by the endogenous overactivity of some aspect of the child's endorphin mechanisms. Some of the earliest symptoms of childhood autism include a lack of crying, a failure to cling to the parents, and a general lack of desire for social interaction. It is as if the child is unable to feel the emotion that is normally generated in social relationships. An insensitivity to pain is another characteristic of both autism and the action of opiate drugs. In addition, opiates can induce stereotyped persistent behaviour that makes learning difficult; again, this is reminiscent of the behaviour of autistic children. Both autistic children and children exposed pre-natally to opiate drugs are often retarded in terms of weight, height, and bone development. The parallels between autism and opiate drugs do not stop here; they both are associated with feeding problems, a readily elicited gag reflex, and a predisposition to epileptic seizure.

Clearly these similarities could be coincidental, but they are sufficiently striking to make worthwhile the exploration of the possibility that a common aetiology exists. The diagnosis of autistic behaviour in humans is difficult, but clearly it is more of a problem to be sure that you are genuinely examining similar mechanisms in both human and non-human animals. However, given that a failure to develop appropriate social bonds is such a prominent characteristic of the autistic child, the study of social behaviour in animals may suggest valuable hypotheses.

EXPERIMENTAL STUDIES OF SOCIAL INTERACTION

The first part of this chapter has argued that powerful evolutionary pressures have bred into humans a tendency to develop several kinds of social bond. It seems very probable that these bonds are mediated by particular circuits in the brain with a distinctive neurochemistry. The making of satisfying social relationships is associated with feelings of well-being, and the breaking of bonds may, in the short-term, be associated with anguish and depression and, in the longer term, with psychiatric problems. Thus, it can be argued that an understanding of the neurochemical basis of social relationships may give valuable insights into the aetiology of various psychiatric complaints, and may offer means of drug screening in this area. It is towards these objectives that the following experiments are tentatively directed.

The social behaviour of adult mice

Both male and female mice were individually housed for a period of 21 days before testing. The isolation of male mice produces animals that are likely to be aggressive; and the isolation of females results in high levels of timid/defensive behaviour. Other male animals remained in groups of six until testing; these were rendered temporarily anosmic by applying, under ether anaesthesia, 25 µl of 4 per cent zinc sulphate to the nasal tract. Anosmic animals spend little time in social investigation and are non-aggressive, thus any social behaviour with which they are involved is likely to be initiated by other animals. The isolated animals were drug-treated and then an anosmic male was introduced. The resulting social behaviour was video-taped for ten minutes. The record was analysed by allocating the time spent displaying one of the following categories of behaviour. The behaviours are described in more detail elsewhere (Benton *et al.* 1983).

1. Non-social behaviour; digging, rearing, self-grooming, walking, and sniffing the cage.

2. Social behaviour; approaching, following, grooming or sniffing the partner.

3. Aggressive behaviour; sideways and upright offensive postures, chasing, lunging at or biting the opponent.

4. Timid/defensive behaviour; sideways and upright offensive postures, avoiding or fleeing from the opponent.

Mu agonists and social interaction

When administered to male mice, neither morphine nor DAGO (Handa *et al.* 1981), both mu agonists, influenced any of the four categories of behaviour (Table 8.1). In contrast, both mu agonists influenced the behaviour of isolated females. Morphine initially decreased the incidence of timid/defensive behaviour (Fig. 8.1), although the other categories of behaviour were unaffected (Benton *et al.* 1985). A similar finding resulted when DAGO was administered to female mice; initially the incidence of timid/defensive behaviour was lower (Benton 1985). In summary, although neither mu agonist influenced the behaviour of aggressive male mice, they both decreased timid and defensive behaviour in females.

Kappa agonists and social interaction

The impact of the kappa agonist U-50488 (Von Voigtlander *et al.* 1983) on the social interaction of male mice was complex (Benton 1985), but could be explained as resulting from an increase in timid and anxious behaviour (Fig. 8.2). Following treatment with the kappa agonist, the mice spent very little time in aggressive interactions; they displayed non-social rather than social behaviour and, when they did explore socially, this was followed by a high incidence of timid/defensive behaviour. These findings are very similar to those obtained following the administration of another kappa agonist, tifluadom (Romer *et al.* 1982), to male mice. They spent less time socially exploring the standard opponent, displayed more timid/defensive behaviour, and were less aggressive (Benton *et al.* 1985). Poshivalov (1986) similarly reported

TABLE 8.1. *The influence of two mu agonists on the social interaction of male mice*

	Social behaviour	Non-social behaviour	Offensive behaviour	Timid/defensive behaviour
Saline	163.70 ± 84.20	394.6 ± 82.40	30.47 ± 37.23	6.39 ± 9.90
1.0 mg/kg Morphine sulphate	127.06 ± 58.74	382.3 ± 64.98	79.51 ± 57.32	2.54 ± 2.32
2.5 mg/kg Morphine	128.44 ± 48.02	369.1 ± 89.51	76.04 ± 78.31	10.45 ± 12.93
	$F_{(2,27)}=0.97$, N.S.	$F_{(2,27)}=0.34$, N.S.	$F_{(2,27)}=2.08$, N.S.	$F_{(2,27)}=1.73$, N.S.
Saline	223.6 ± 90.31	342.0 ± 89.1	30.2 ± 16.72	6.9 ± 4.50
250 µg/kg DAGO	192.4 ± 79.92	336.8 ± 87.8	56.4 ± 30.47	4.4 ± 2.71
1.0 mg/kg DAGO	196.5 ± 80.18	345.4 ± 92.1	49.9 ± 26.33	4.3 ± 3.12
	$F_{(2,27)}=0.80$, N.S.	$F_{(2,27)}=0.05$, N.S.	$F_{(2,27)}=1.39$, N.S.	$F_{(2,27)}=1.24$, N.S.

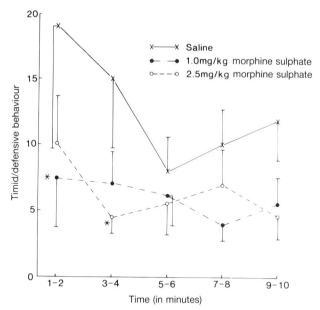

FIG. 8.1. The influence of morphine sulphate on the timidity of female mice. At two points, using Dunnett's test, there was a significant difference between the morphine-treated animals and the controls. *$p < 0.05$.

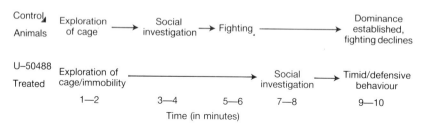

FIG. 8.2. A summary of the influence of U-50488 on the behaviour of male mice. Mice treated with the kappa agonist, U-50488, were more defensive; they socially-investigated at a later stage, and they displayed timid behaviour.

that the kappa agonist bremazocine decreased aggression and active forms of defence, while they increased passive forms of defence. In contrast, in female mice neither tifluadom nor U-50488 influenced any of the three categories of behaviour (Table 8.2).

The testing of a model

Figure 8.3 restates the above results, describing on a continuum the

TABLE 8.2. *The influence of two kappa agonists on the social interaction of female mice*

	Social behaviour	Non-social behaviour	Offensive behaviour	Timid/defensive behaviour
Saline	121.57 ± 54.76	340.39 ± 93.66	0 (0–0.5)	137.73 ± 119.25
0.5 mg/kg Tifluadom	120.91 ± 94.14	398.72 ± 78.57	0 (0–3)	79.96 ± 84.41
1.0 mg/kg Tifluadom	94.14 ± 73.04	437.27 ± 100.01	0 (0–3.9)	68.04 ± 114.19
	$F_{(2,24)}=0.49$, N.S.	$F_{(2,24)}=2.57$, N.S.	N.S.	$F_{(2,24)}=1.09$, N.S.
Saline	166.09 ± 52.73	350.54 ± 66.87		81.65 ± 51.48
2.5 mg/kg U-50488	160.53 ± 51.69	357.55 ± 70.31		81.97 ± 55.73
5.0 mg/kg U-50488	178.48 ± 91.53	333.57 ± 71.28		87.47 ± 62.32
	$F_{(2,27)}=0.15$, N.S.	$F_{(2,27)}=0.28$, N.S.		$F_{(2,27)}=0.04$, N.S.

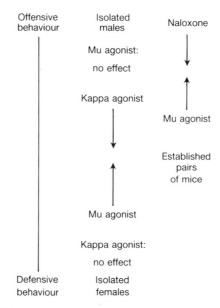

FIG. 8.3. A model and some predictions. The left-hand column describes the ways in which animals can interact on meeting. The middle column suggests that mu agonists push animals up, and kappa agonists down, the continuum. The right-hand column represents two predictions from this model.

possible ways in which animals can interact when they meet socially. At one extreme, some animals are very likely to initiate an overtly aggressive encounter; this is true of many isolated male mice. At the other extreme, some animals are predisposed to act submissively, something true of isolated female mice. Naturally most animals, most of the time, will lie between these two poles. The above data can be explained by assuming that mu agonists push mice up the continuum and kappa agonists have the opposite effect, moving them downwards.

When a male mouse reacts to isolation by displaying many of the characteristics of a dominant, territory-holding mouse (Benton and Brain 1979), it is as if he is towards the top of the continuum. Thus the administration of mu agonists should not be influential (Table 8.1) as the mu receptors are already being stimulated by endogeneous peptides. Kappa drugs should, however, have an influence as they are predicted to decrease aggressive inclinations (Fig. 8.2). An exactly opposite picture results with an isolated female mouse that often displays defensive behaviour, even when the intruding male ignores her. In this case it is as if kappa mechanisms are already active and thus kappa agonists are without influence (Table 8.2). In contrast mu agonists should decrease the timidity of isolated females (Fig. 8.1). These data are summarized in the middle column of Figure 8.3.

When Figure 8.3 was first drawn, it was merely a *'post hoc'* summary of experimental findings. As such its value would be greatly increased if in addition it could be shown to have predictive value. Two predictions are illustrated in the right hand column of Figure 8.3. In doses much lower than often used in behavioural experiments, naloxone binds preferentially to mu sites, and is capable of blocking the mu-mediated pain relieving effects of morphine, by binding preferentially to mu sites. The model predicts that low doses of naloxone should have an anti-aggressive influence. Benton (1984) found exactly that, a low dose (0.1 mg/kg) of naloxone had a potent anti-aggressive influence.

Another prediction is that mice that are in the middle of the continuum, and are not particularly disposed to display overt aggression, but are capable of doing so should the occasion demand it, should react to the stimulation of the mu site by displaying aggressive behaviour. Figure 8.4 reports a test of this second prediction. Pairs of mice were allowed to live together for one week during which they developed a dominant/subordinate relationship. Once such a relationship is established, in the majority of cases, overt fighting is rarely if ever observed. A low dose of morphine very potently increased fighting during a ten-minute observation period. The higher dose, one that in previous studies would have been a very low one, was without effect, probably reflecting a sedative influence. This aggression-enhancing influence of morphine is similar to the report of Poshivalov (personal communication, 1985) that another mu agonist, fentanyl, similarly stimulated aggression.

In summary, the reported data can be explained by assuming that mu and kappa mechanisms have opposite influences on social interactions in mice. The stimulation of mu sites increases the likelihood of offensive behaviour, and decreases the likelihood of timid/submissive behaviour.

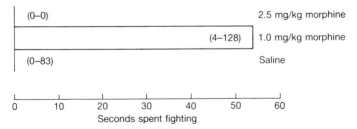

FIG. 8.4. The influence of morphine on the behaviour of established pairs of male mice. Pairs of male mice lived together for a week. They were then injected with morphine, and the time spend fighting was recorded (medians and ranges). A low dose of morphine potently stimulated fighting. $p < 0.001$.

The opposite picture results from the administration of kappa agonists, the likelihood of defensive behaviour is increased, and offensive behaviour is decreased.

Discussion

At first it may appear that this model conflicts with much previous work. It has been known for many years that acute administration of narcotic drugs suppresses aggression produced by many procedures and in several species. It is not surprising that morphine, a potent analgesic, in rats (Lal *et al.* 1975) and monkeys (Emley *et al.* 1970) blocks aggression induced by electric shocks. However, morphine also reduces aggression in situations that do not involve pain; there are reports that it suppresses isolation-induced fighting in mice (DaVanzo *et al.* 1966; Janssen *et al.* 1960) and mouse-killing by rats (Janssen *et al.* 1962). In addicted animals the withdrawal of morphine is associated with spontaneously-violent aggression (Thor *et al.* 1970).

It seems very likely that these early studies of the influence of morphine on aggression are not mediated solely via opiate sites. In fact it has been hypothesized that aggression following morphine withdrawal reflects a supersensitivity of dopamine receptors. In addition, a role for serotonergic and cholinergic mechanisms have been suggested (Gianutsos and Lal 1978). It is relevant to contrast the mu-mediated analgesic influence of morphine (E-D-50, 2.3 mg/kg) with the doses used in these early studies. Animals addicted to morphine may receive hundreds of milligrams a day; DaVanzo and his colleagues (1966) reported that the E-D-50 for the inhibitory influence on morphine on isolation-induced aggression was 21.2 mg/kg. Such doses are certain to have non-specific effects. It is hardly surprising that the low doses of morphine used in the present studies have had a different effect.

THE INFLUENCE OF DRUGS ACTING AT OPIATE SITES ON MURINE MATERNAL BEHAVIOUR

The present studies attempt for the first time to distinguish the roles of mu and kappa opiate receptors in maternal behaviour.

The retrieval of young by lactating females

The presence of a pup outside the nest is a very powerful stimulus for a mother mouse. Very rapidly she picks up the pup and returns it to the nest; it seems almost compulsive, and will be repeated over and over again. In the present study, three days after birth mothers were taken from the nest and treated with either morphine, U-50488, or an opioid antagonist. After ten minutes, six of the litter were taken from the nest and placed at the far end of the cage. The mother was returned to the home cage and the time taken for her to return each of the young to the nest was recorded. In case a desire to explore initially predominated on return to the cage, after five minutes six of the young were again removed from the nest and the time to retrieve them was recorded.

Figure 8.5 shows the influence of morphine on retrieval of pups under these conditions. It is apparent that morphine greatly disrupted this behaviour. In contrast, the kappa agonist U-50488 did not influence the retrieval of the young. Of the three opioid antagonists examined in this test, the kappa antagonist Mr 2266 was without effect; both naloxone (in a low dose that was probably acting mainly at mu sites) and the lower dose of the delta antagonist ICI 154,129 (Shaw *et al.* 1982) inhibited retrieval during the first block of trials (Figure 8.6).

The interest of virgin females in pups
As the mothers may have learnt to respond to the young by the third day, a second situation was examined in which maternal drive and experience were absent. Six-week-old virgin females were paired for a week before testing. During behavioural testing one was removed from the home cage for the duration of the experiment. The other mouse was injected and returned to the home cage for ten minutes to allow the drug to act. A three-day-old pup was then taken from the nest and placed in the cage of the drug-treated animal for a period of five minutes. In the AP strain used in these experiments a pup is a powerful stimulus; in the most extreme cases a nest was built, the pup was retrieved, licked, and the lactation position was taken up. Using an electronic timer the time spent displaying various maternally-related behaviours was recorded. Only three are presently reported, as the time spent

(1) licking; the pup was washed using the tongue, often being moved by the adult so the body was systematically cleaned;

(2) sniffing the pup;

(3) rearing; standing on the rear legs to explore the cage (this measure is sensitive to both sedative and stimulatory drug influences).

Figure 8.7 illustrates the impact of morphine on the interest of virgin females in pups. The time spent licking the young was significantly decreased. This was a specific effect; all the adult females initially explored the pups but, when treated with morphine, this was less likely to lead to maternal licking. Equally the time spent rearing was not affected. In contrast, U-50488, a kappa agonist, produced a more marked

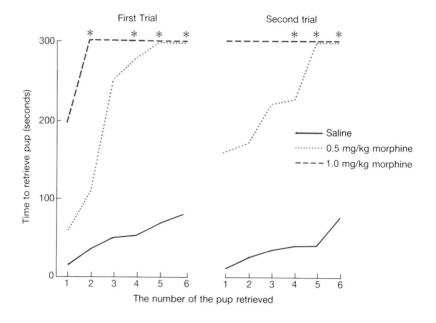

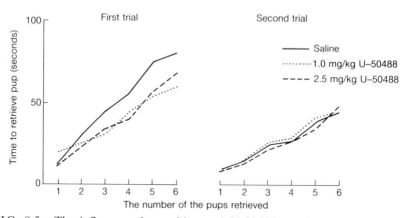

FIG. 8.5. The influence of morphine and U-50488 on the time taken by a mother to retrieve mouse pups. Each point was analysed using Kruskal-Wallis one-way analysis of variance. *p < 0.05 or better.

disruption of behaviour. The licking of the young was lower in the drug-treated animals (although this may be secondary to such other influences) as U-50488 increased rearing, and the time spent exploring the cage may have resulted in less time to either sniff or lick the pup.

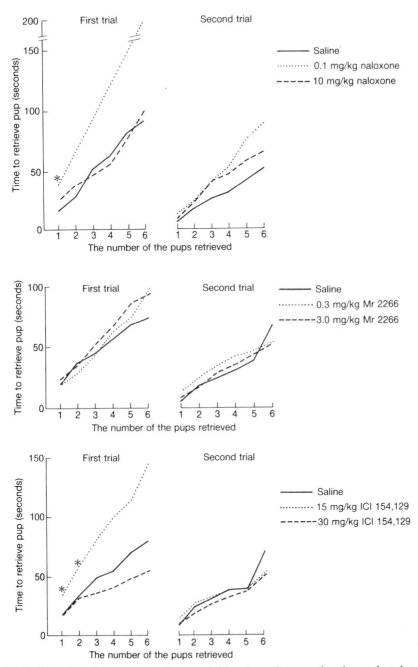

FIG. 8.6. The influence of three opioid antagonists on the time taken by a mother to retrieve mouse pups. Each point was analysed using Kruskal–Wallis one-way analysis of variance. *p < 0.05.

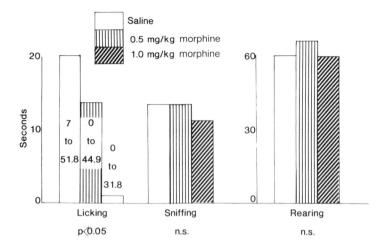

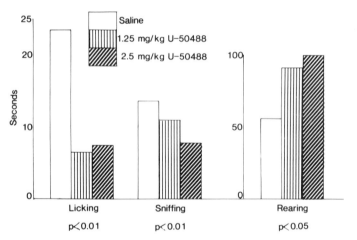

FIG. 8.7. The interest of virgin female mice in a pup, after injection with morphine (top) or U-50488 (bottom). During a five-minute test, the time spent examining a pup (licking or sniffing) or exploring the cage (rearing) was recorded.

Discussion

These data can be viewed as being consistent with the speculation of Grimm and Bridges (1983) that the high levels of beta-endorphin, produced during pregnancy, inhibit maternal behaviour. They found similar results to the present study in the rat; morphine prevented the onset of maternal behaviour. As U-50488 was without effect, it appears that mu, but not kappa, mechanisms are implicated. As the administration of selective antagonists resulted in relatively little disruption of maternal behaviour, this suggests that endorphins were not being released postnatally under the conditions of this test. The initial inhibition of maternal behaviour by both naloxone and ICI 154,129 probably reflects other mechanisms; if these antagonists had been influencing maternal motivation, they should have had an impact in the second test block when, for example, morphine was still active. It is perhaps relevant that naloxone has been reported to influence the response to novelty, and its only significant impact was when the mother first returned to the cage. ICI 154,129 has also been reported to influence emotionality (Benton *et al.* 1984); this may again be the origin of the initial disruption of pup retrieval. The results of studying virgin females can be viewed to be consistent with the pup retrieval findings. Morphine produced a specific effect; it appeared that again it was an interest in the young that was being specifically influenced. Although the kappa agonist similarly decreased interest in the young, the profile of changes suggested a much more general change in behaviour, of which decreased licking of the young was secondary to other changes.

THE INFLUENCE OF OPIOID DRUGS ON THE ULTRASONIC CALLING OF MOUSE PUPS

Five-day-old pups were injected IP with one of various opioid drugs, and were then returned to the nest for ten minutes to allow the drug to act. They were then placed in a 5 cm-diameter glass beaker suspended so that the bottom 4 cm was in a water bath maintained at either 20°C or 37°C. For a period of 21 minutes the ultrasonic calls of each pup were monitored automatically. A microphone placed over the beaker directed the calls through electronic filters that responded to calls from 55 to 80 Hz. Every 14 msec a microcomputer recorded the presence or absence of a call, and the resulting summated scores are the data reported. The mice were kept on a reversed-lighting schedule (lights on 18.30–06.30), and were tested during the last two-thirds of the dark half of the cycle.

The data obtained at 20°C following the injection of either morphine or U-50488 are reported in Figure 8.8. Following morphine administration, there was a marked decrease in vocalization ($F(2,18)=7.96$, $p < 0.01$). Similarly U-50488 potently decreased vocalization ($F(2,18)=8.09$, $p < 0.01$). Figure 8.8 shows a clear decline in the number of ultrasonic calls over the duration of the experiment. It seemed possible that the gradual cooling of the pup's body during the experiment may have stimulated endorphin release that lead to the decline in the rate

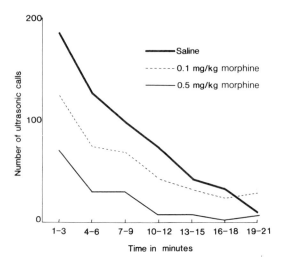

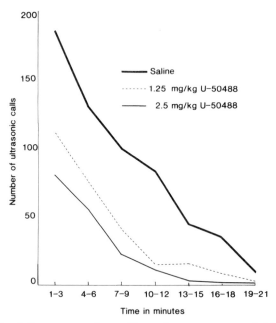

FIG. 8.8.   The influence on the calling of mouse pups of 20°C of opioid agonists, morphine (top) and U-50488 (bottom).

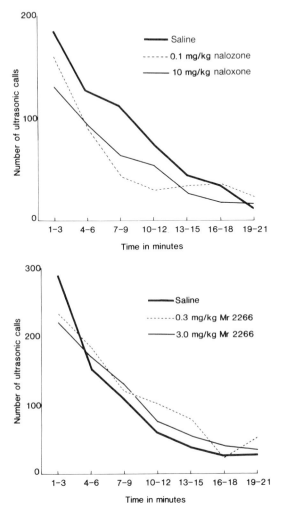

FIG. 8.9. The influence on the calling of mouse pups at 20°C of opioid antagonists, naloxone (top) and Mr 2266 (bottom).

of calling. It was argued that, if this were the case, opioid antagonists would prevent this change over time. The hypothesis was not supported; in the case of naloxone, neither the drug main effect (F(2,18)=0.72, n.s.) nor the drug/time interaction (F(12,108)=0.89, n.s.) reached statistical significance (Fig. 8.9). Similarly, with the kappa agonist Mr 2266, both the drug main effect (F(2,16)=0.45) and interaction with time (F(12,90)=1.44) were statistically insignificant. Thus there was no evidence that cooling the pups stimulated endorphin release.

A similar argument could be applied to pups kept at 37°C. As pups

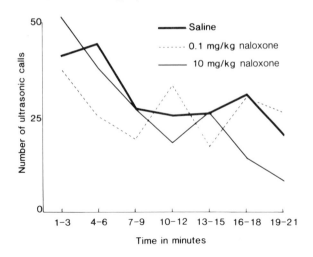

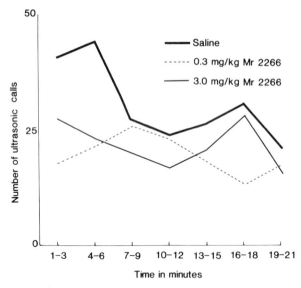

FIG. 8.10. The influence on the calling of mouse pups at 37°C of opioid antagonists, naloxone (top) and Mr 2266 (bottom).

kept at nest temperature call relatively infrequently, it seemed possible that endorphin release was inhibiting vocalization. If so, then opioid antagonists could be expected to stimulate the rate of calling. The impact of naloxone on the rate of calling of pups kept at 37°C is reported in Figure 8.10. Again, neither the drug main effect ($F(2,18)=0.14$) nor the interaction with time ($F(12,108)=0.89$) reached statistical significance.

Similarly, Mr 2266 did not influence the rate of calling (main effect $F(2,18)=1.91$, interaction with time $F(12,108)=0.90$).

Discussion

The present data confirm, in mice, that morphine can decrease the vocalization of the young, when separated from the mother. Previously this observation has been made in puppies and chicks (Panksepp *et al.* 1978), and guinea pigs (Herman and Panksepp 1978). In all these species, morphine influences vocalization in very low doses (0.1 to 0.5 mg/kg). In fact, separation-induced vocalization is probably the behaviour most susceptible to morphine. The only systematic study of pharmacological influences on separation vocalization is in the chick. Panksepp and his colleagues (1980) examined the major drugs used therapeutically in psychiatry, as well as those acting at cholinergic, noradrenergic, dopaminergic and serotonergic sites. Only clonidine, an adrenergic agonist, approached the potency of morphine. The finding that a kappa agonist inhibits vocalization (Fig. 8.8) is the first examination of this question; it appears that both mu and kappa mechanisms are involved in this behaviour.

Panksepp and his colleagues (1985) argued that endogenous opioids inhibit the brain circuits that produce separation-induced vocalization. A central prediction from this viewpoint is that opioid antagonists should increase calling. Although this has been reported in guinea pigs (Herman and Panksepp 1978), chicks (Panksepp *et al.* 1980), and mice (Robinson *et al.* 1985), it is not a universally-observed finding. Panksepp and his colleagues (1985) noted that, in the chick, the effect is influenced by circadian variables; it is observed during the day when the bird is awake, rather than the night. A low baseline of calling was again more likely to result in a naloxone-induced potentiation. Given the difficulty of observing an influence of naloxone in other species, it is perhaps not surprising that the present study has similarly failed to find a significant influence. Whether a systematic variation of the time of day and other variables will result in conditions under which naloxone stimulates vocalization in the mouse pup awaits further work.

Lehr (1986) has proposed that the calling of chicks be used to screen anti-depressant drugs. When isolated the initial rate of calling is high, the stage of protest; later, the bird enters the stage of resignation when the rate of calling decreases. Lehr found that anti-depressant reference drugs resulted in a high rate of calling at the time when the controls were becoming resigned. In contrast, drugs without anti-depressant actions either did not influence the rate of calling or inhibited it even further. This observation gives further support for the theoretical position, discussed above, that separation calling may reflect mechanisms associated with human depression.

CONCLUSION

Social attraction and bonding are essential aspects of human evolutionary development; for example, parent–infant attraction is necessary during a prolonged pre-adult stage when cultural, linguistic, and manual skills develop. In addition, the importance of a warm, lasting, and emotionally-satisfying bond between infant and parent has been increasingly recognized as important for the development of a healthy adult personality. As it seems very probable that several kinds of social bond are mediated by brain circuits that developed early in our evolution, and that could be expected to have a distinct neurochemistry, it is strange that this area of study has been ignored to a large extent. In a series of studies, the role of various opiate receptors in the social behaviour of mice has been examined.

The view that opioid drugs influence a wide range of social behaviour is given general support by these experimental findings. The extent to which, in some cases, the results may reflect a pharmacological influence, rather than a physiological mechanism, is a matter for debate. It may be expected that, if a particular drug were influencing endogenous mechanisms, small doses, that bind preferentially to particular sites should be effective. The very low doses of morphine used in these studies suggest that it is stimulating a specific receptor and an endogeneous mechanism.

In psychopharmacological studies, it is generally easier to disrupt than to stimulate behaviour; disruption may often reflect any of a range of non-specific effects. It is therefore relevant that much of the present data reflects a suppression of behaviour. The case for the existence of an underlying endorphin-mediated mechanism would be greatly increased if particular pairs of agonists and antagonists could be shown to have opposite influences. In several cases this finding, although looked for, has not been obtained. It is, however, always possible that negative results reflect inappropriately-chosen experimental parameters. In other situations, for example vocalization in the chick (Panksepp *et al.* 1985) and pain sensitivity (Yaksh *et al.* 1976), it has proved difficult to demonstrate an influence of naloxone.

The impact of opioid antagonists will only be demonstrable if conditions are such that endorphins are being released. If conditions are such that either minimum or maximal responses are already being elicited, then drug-induced decreases or increases will be impossible to demonstrate. In situations such as the mother retrieving the young, the behaviour is so urgent that it is perhaps unreasonable to expect to observe a potentiation.

The social interaction of adult mice is the model for which the most convincing case of endorphin involvement can be made. In the studies reported above, both agonists and antagonists influenced offensive and defensive behaviour in a consistent manner; in addition a model resulted (Figure 8.3) that was able to predict experimental findings. It seems fair to conclude that mu agonists and kappa agonists have opposite influences on offensive and timid/defensive behaviour.

The interest of female mice in pups was decreased by the mu agonist, morphine, but not by the kappa agonist, U-50488; these findings were suggested as being consistent with the high level of beta-endrophin produced during pregnancy (which has, as one function, the inhibition of maternal behaviour). The use of both mu and kappa agonists inhibited the vocalization of pups when separated from the mother.

These findings emphasize the need in future work to further distinguish the roles of the various opiate receptors. It is clear that the days are gone when large doses of drugs that block mu, kappa, and delta sites (such as naloxone), can be used to merely conclude that endorphins are involved in a particular mechanism. The possibility must be explored that there are several opioid mechanisms that may or may not influence a particular behaviour in similar ways.

The general proposition, that an understanding of the neurochemical basis of social bonding may lead to an insight into the aetiology of various psychiatric disorders, and the development of novel pharmacological treatments, seem worthy of further examination. The clear evolutionary importance of the development of various forms of social bonding, and the widespread problems associated with their disruption, give strong support to this line of argument. To date the hypothesis, that endorphins are intimately associated with a range of social behaviours, seems the most promising neurochemical hypothesis, although it is unlikely that other aspects of the brain's chemistry will not be implicated in these types of behaviour.

REFERENCES

Adam, K. S. (1982). Loss, suicide and attachment. In *The place of attachment in human behavior* (eds. C. Murray Parkes and J. Stevenson-Hinde), pp. 269–94. Tavistock Publications, London.

Alleva, E., Caprioli, A., and Laviola, G. (1986). Postnatal social environmental affects morphine analgesia in male mice. *Physiology and Behavior* **36**, 779–81.

Argyle, M. (1986). *The psychology of happiness.* Methuen, London.

Benton, D. (1984). The long-term effects of naloxone, dibutyryl cyclic CMP and chlorpromazine on aggression in mice monitored by an automatic device. *Aggressive Behavior* **10**, 79–89.

—— (1985). Mu and kappa receptor involvement in agonistic behaviour in mice. *Pharmacology, Biochemistry and Behavior* **23**, 871–6.

—— and Brain, P. F. (1979). Behavioural comparisons of isolated dominant and subordinate mice. *Behavioral Processes* **4**, 211–19.

——, Brain, P. F., Jones, S., Colebrook, E., and Grimm, V. (1983). Behavioural examinations of the anti-aggressive drug fluprazine. *Behavioral Brain Research* **10**, 325–38.

——, Brain, S. and Brain, P. F. (1984*a*). Comparison of the influence of the opiate delta receptor antagonist ICI 154,129 and naloxone on social interaction and behaviour in an open field. *Neuropharmacology* **23**, 13–17.

——, Dalrymple-Alford, J. C., McAllister, K. H., Brain, P. F., and Brain, S. (1984*b*). Comparison in the mouse of the effect of the opiate delta receptor antagonist ICI 154,129 and naloxone in tests of extinction, passive avoidance and food intake. *Psychopharmacology* **82**, 41–5.

——, Smoothy, R. and Brain, P. F. (1985). Comparisons of the influence of morphine sulphate, morphine-3-glucuronide and tifluadom on social encounters in mice. *Physiology and Behavior* **35**, 689–93.

Birtchnell, J. (1975). The personality characteristics of early-bereaved psychiatric patients. *Social Psychiatry* **10**, 97–103.

Bowlby, J. (1965). *Child care and the growth of love.* Pelican Books, Harmondsworth.

—— (1980). *Loss.* Hogarth Press, London.

Bridges, R. S. and Grimm, C. T. (1982). Reversal of morphine disruption of maternal behavior by concurrent treatment with opiate antagonist naloxone. *Science* **218**, 166–8.

Brown, G. W. (1982). Early loss and depression. In *The place of attachment in human behavior* (eds. C. Murray Parkes and J. Stevenson-Hinde), pp. 232–68. Tavistock Publications, London.

Darwin, C. (1872). *The expression of emotions in man and animals.* University of Chicago Press, Chicago.

DaVanzo, J. P., Daugherty, M., Ruckart, R., and Kang, L. (1966). Pharmacological and biochemical studies in isolation-induced fighting mice. *Psychopharmacologia* **9**, 210–19.

Emley, G. S., Hutchinson, R. R., and Brannan, I. B. (1970). Aggression: effects of acute and chronic morphine. *Michigan Mental Health Research Bull.* **4**, 23–6.

Fontana, V. J. (1973). *Somewhere a child is crying: Maltreatment—causes and prevention.* Macmillan, New York.

Gianutsos, G. and Lal, H. (1978). Narcotic analgesics and aggression. *Modern Problems in Pharmacopsychology* **13**, 114–38.

Grimm, C. T. and Bridges, R. S. (1983). Opiate regulation of maternal behavior in the rat. *Pharmacology, Biochemistry and Behavior* **19**, 609–16.

Handa, B. A., Lane, A. C., Lord, J. A. H., Morgan, B. A., Rance, M. J., and Smith, C. F. C. (1981). Analogues of β-LPH 61-64 possessing selective agonist activity at μ-opiate receptors. *European J. Pharmacol.* **70**, 531–40.

Hard, E. and Engel, J. (1986). Effects of a putative 5-HT agonist, 8-OH-DPAT on emotional reactivity in postnatal rats. *Psychopharmacology* **89**, S49.

Henderson, S. (1982). The significance of social relationships in the etiology of neurosis. In *The place of attachment in human behavior* (eds. C. Murray Parkes and J. Stevenson-Hinde), pp. 205–31. Tavistock Publications, London.

Herman, B. H. and Panksepp, J. (1978). Effects of morphine and attachment on separation distress and approach attachment: evidence for opiate mediation of social affect. *Pharmacology, Biochemistry and Behavior* **9**, 213–20.

Hofer, M. A. (1984). Relationships as regulators: a psychobiologic perspective on bereavement. *Psychosomatic Medicine* **46**, 183–97.

Janssen, P. A. J., Jageeneau, A. H., and Niemeggers, C. J. E. (1960). Effects of various drugs on isolation-induced fighting behaviours of male mice. *J. Pharmacol. and Exper. Ther.* **129**, 471–5.

——, Niemegeers, C. J. E., and Verbruggen, F. J. (1962). A propos d'une methode d'investigation de substances susceptibles de modifier le comportement agressif inné du rat blanc vis-à-vis de la souris blanche. *Psychopharmacologia* **3**, 114–23.

Kalat, J. W. (1978). Speculations on similarities between autism and opiate addiction. *J. Autism and Childhood Schizophrenia* **8**, 477–9.

Kostowski, W., Czlonkowski, A., Reverski, W., and Piechocki, T. (1977). Morphine action in group-housed and isolated rats and mice. *Psychopharmacology* **53**, 191–3.

Lal, H., Wauquier, A., and Niemegeers, C. J. E. (1975). Characteristics of narcotic dependence by oral ingestion of fentanyl solution. *Pharmacologist* **17**, 237.

Lehr, E. (1986). A new behavioural model for antidepressants. *Meeting of the European Behavioural Pharmacology Society*, Antwerp, July 2–6.

Panksepp, J. (1979). A neurochemical theory of autism. *Trends in Neuroscience* **2**, 174–7.

—— (1980). Brief social separation: Pain responsivity and morphine analgesia in young rats. *Psychopharmacology* **72**, 111–12.

——, Herman, B. H., Vilberg, T., Bishop, P., and De Eskinazi, F. G. (1978). Endogenous opioids and social behavior. *Neuroscience and Biobehavioral Review* **4**, 473–87.

——, Najam, N., and Soares, F. (1979). Morphine reduces social cohesion in rats. *Pharmacology, Biochemistry and Behavior* **11**, 131–4.

——, Meeker, R., and Bean, N. J. (1980). The neurochemical control of crying. *Pharmacology, Biochemistry and Behavior* **12**, 437–43.

——, Siviy, S. M., and Normansell, L. A. (1985). Brain opioids and social emotions. In *The psychobiology of attachment and separation* (eds. M. Reite and T. Field), pp. 3–49. Academic Press, New York.

Passingham, R. (1982). *The human primate*. Freemans, San Francisco.

Petkov, V. V., Konstantinova, E., and Grachovska, T. (1985). Changes in brain opiate receptors in rats with isolation syndrome. *Pharmacological Research Communications* **17**, 575–84.

Poshivalov, V. P. (1986). Ethological pharmacology as a tool for animal aggression research. In *Cross disciplinary studies on aggression* (eds. P. F. Brain and J. M. Ramirez), pp. 96–114. Publicaciones de la Universidad de Sevilla, Sevilla.

Robinson, D. J., D'Udine, B., and Olivero, A. (1985). Naloxone influences ultrasonic calling in young mice. *Behavioral Processes* **11**, 253–5.

Romer, D., Buscher, H. H., Hill, R. C., Mauer, R., Petcher, T. J., Zeugner, H., Benson, W., Finner, E., Milkowski, W., and Thies, P. W. (1982). An opioid benzodiazepine. *Nature* **298**, 759–60.

Rossi, J., Sahley, T. L., and Panksepp, J. (1983). The role of brain norepinephrine in clonidine suppression of isolation-induced distress in the domestic chick. *Psychopharmacologia* **79**, 338–42.

Shaw, J. S., Miller, L., Turnbull, M. J., Gormley, J. J., and Morley, J. S. (1982). Selective antagonists at the opiate delta-receptor. *Life Science* **31**, 1263–6.

Thor, D. H., Hoates, D. L., and Thor, C. J. (1970). Morphine induced fighting and prior social experience. *Psychonomic Science* **18**, 137–9.

Von Voigtlander, P. F., Lahti, R. A., and Ludens, J. H. (1983). A selective and structurally novel non-mu (kappa) opioid agonist. *J. Pharmacol. and Exper. Ther.* **224**, 7–12.

Yaksh, T. L., Yeung, J. C., and Rudy, T. A. (1976). An inability to antagonize with naloxone the elevated nociceptive thresholds resulting from electrical stimulation of the mesencephalic central gray. *Life Sciences* **18**, 1193–8.

9

Caffeine physical dependence and reinforcement in humans and laboratory animals

ROLAND R. GRIFFITHS and PHILLIP P. WOODSON

INTRODUCTION

Caffeine (1,3,7-trimethylxanthine, Fig. 9.1) is the most widely-used behaviourally active drug in the world (Gilbert 1984), with more than 80 per cent of adults in North America regularly consuming caffeine (Gilbert 1976; Graham 1978). In the United States and Canada, daily per capita caffeine consumption has been estimated to be 211 and 238 mg, respectively. These figures are about half those estimated for the United Kingdom (444 mg) and Sweden (425 mg), which are particularly heavy tea- and coffee-consuming countries, respectively (Gilbert 1984).

Sixty-three species of caffeine-containing plants have been identified, with the most widely-used sources of dietary caffeine being coffee, tea, maté, kola, cocao pod, and guarana paste (Gilbert 1984). Caffeine was, of course, first used by societies which had ready access to the naturally-occurring caffeine-containing plants, and records of caffeine use date back at least 1600 years (Graham 1984). Use of these caffeine-containing foods spread systematically, despite recurring efforts to restrict or eliminate their use (Austin 1979; Graham 1984).

Perhaps it should not be surprising that, given the high prevalence and persistence of caffeine use, caffeine has been periodically identified as a drug of abuse (Gilbert 1976; Austin 1979; Greden 1981). Drugs of abuse have two major characteristics;

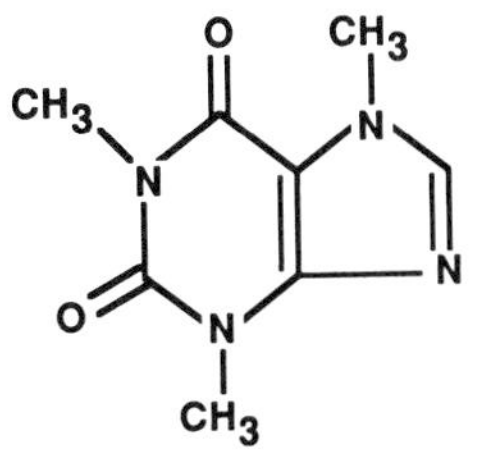

FIG. 9.1. Chemical structure of caffeine.

141

(1) they produce adverse effects (they have the capacity to harm the individual and/or society), and

(2) they have reinforcing properties (Griffiths *et al.* 1985).

The purpose of this chapter is to evaluate the current understanding of the physical dependence-producing and behaviour-reinforcing effects of caffeine.

PHYSICAL DEPENDENCE IN LABORATORY ANIMALS

Among the most clearly documented adverse effects of habitual caffeine use is physical dependence. Physical dependence is manifested by biochemical, physiological, and/or behavioural disruptions that occur upon termination of chronic drug administration. Physical dependence may contribute to the abuse liability of a drug in two ways;

(1) as an adverse effect of drug use that is revealed upon discontinuation of drug use; and

(2) as a potential mechanism by which the reinforcing effects of a drug may be enhanced.

Despite the fact that reliable methods have been developed for evaluating the physical dependence potential of various drug classes in laboratory animals (Martin 1977; Brady and Lukas 1984), few studies have been conducted with caffeine. This literature has been reviewed recently (Griffiths and Woodson, in press, *a*). Six reports have been published using rats, most of which document substantial behavioural disruptions following cessation of chronic caffeine dosing (Boyd *et al.* 1965; Vitiello and Woods 1977; Carney 1982; Holtzman 1983; Finn and Holtzman 1986; Holtzman and Finn, in press). The most reliable caffeine-withdrawal effect documented to date has been that of decreased locomotor activity. Disruption of operant schedule-controlled behaviour during caffeine withdrawal has been less reliably demonstrated (Carney 1982; Holtzman and Finn, in press). One final pre-clinical study used a taste-aversion paradigm to provide evidence for the aversive properties of absence of caffeine in rats repeatedly exposed to caffeine (Vitiello and Woods 1977).

PHYSICAL DEPENDENCE IN HUMANS

As in research with laboratory animals, experimental methods have been established for evaluating the physical dependence potential of drugs in humans (Martin 1977; Brady and Lukas 1984; Petursson and Lader 1984). In contrast to the relatively few studies documenting physical dependence with caffeine in laboratory animals, physical dependence has been clearly and repeatedly documented in humans.

Thirty-seven published reports, including clinical observations, case studies, survey studies, and experimental manipulations, which provide information about the signs, symptoms, and time course of the caffeine-withdrawal syndrome, have been reviewed recently (Griffiths and Woodson, in press, *a*). Headache and fatigue are the most prominent caffeine-withdrawal symptoms, with a wide variety of other signs and symptoms having been reported at lower frequency (e.g. anxiety, rhinorrhoea, irritability, impaired psychomotor performance, nausea/vomiting, yawning, insomnia, confusion, diaphoresis, muscle pains/stiffness, coffee craving).

When signs and/or symptoms of caffeine withdrawal occur, the severity can vary from mild to extreme. At its worst, caffeine withdrawal has been repeatedly documented to be incompatible with normal functioning, and sometimes totally incapacitating (Dreisbach and Pfeiffer 1943; Goldstein and Kaizer 1969; Greden *et al.* 1980; Cobbs 1982; Rainey 1985). The withdrawal syndrome follows an orderly time-course, with onset occurring at 12 to 24 hours, peak occurring at 20 to 48 hours, and duration most often being described as being about 1 week (Dreisbach and Pfeiffer 1943; Greden *et al.* 1980; Griffiths *et al.* 1986*a*; Wilkin 1986).

As reviewed by Griffiths and Woodson (in press, *a*), the pharmacological specificity of caffeine withdrawal has been established by the observations that;

(1) withdrawal severity is an increasing function of maintenance dose;

(2) withdrawal occurs after administration of caffeine in capsules as well as in beverages;

(3) caffeine-withdrawal symptoms are suppressed by administration of caffeine in capsules, tablets, or beverages;

(4) magnitude of suppression is an increasing function of dose; and

(5) caffeine is more effective at suppressing withdrawal than a variety of other drugs.

The proportion of heavy caffeine users (≥ 500 mg/day) who will experience symptoms after caffeine abstinence is unclear; however, it has been reported to be greater than 80 per cent in a study using a relatively unselected subject population (Dreisbach and Pfeiffer 1943). There has been wide variability in incidence, severity, and duration of withdrawal symptoms, perhaps reflecting stable individual differences.

Although the minimum conditions necessary to induce physical dependence are also unclear, there is evidence for withdrawal symptoms occurring after termination of caffeine after short-term exposure to high doses (≥ 600 mg/day for 6 to 15 days), or after long-term exposure to relatively low doses (two to four cups of coffee/day; Griffiths and Woodson, in press, *a*). Several studies suggest that physical dependence may substantially potentiate the reinforcing effects of caffeine (Goldstein *et al.* 1969; Griffiths *et al.* 1986*a*).

In heavy caffeine users, therapeutic detoxification should be individually tailored by first attempting abrupt cessation, and subsequently using the more involved procedures of caffeine-replacement therapy, supplemental medication, and/or structured behavioural dose-tapering programmes if significant withdrawal symptoms develop during abrupt abstinence (Greden 1981). As with other abused drugs (e.g. heroin, alcohol, and nicotine), patients with histories of heavy caffeine use may be resistant to detoxification, and relapse can be expected to occur frequently (Greden *et al.* 1980; Greden 1981; Wilkin 1986).

One of the most complete characterizations of the behavioural and subjective aspects of the caffeine-withdrawal syndrome in humans was provided in an experiment in which seven subjects with histories of heavy coffee drinking were switched abruptly, under double-blind conditions, from caffeinated coffee to decaffeinated coffee for ten or more days (Griffiths *et al.* 1986*a*). Mean caffeine intake preceding the withdrawal phase was 1.25 g/day. The withdrawal phase was the first occasion during their experimental participation when subjects were exposed to decaffeinated coffee for more than a 24-hour period. As described in more detail below, substitution of decaffeinated coffee did not significantly affect number of cups of coffee consumed, but was associated with transient decreases in subject ratings of coffee liking. The results of the experiment also showed that substitution of decaffeinated coffee produced an orderly withdrawal syndrome which peaked on Day 1 or Day 2 after substitution, and then decreased progressively over the next five or six days. The withdrawal syndrome, which was detectable on subject-rated, staff-rated, and objective behavioural measures, was characterized by increased headache, sleepiness and laziness, and decreased alertness and activeness.

In this study the most sensitive and reliable subject-rated withdrawal symptom was headache; this remained significantly elevated over prewithdrawal levels on the third day after substitution of decaffeinated coffee. Figure 9.2 presents detailed information about reports of headache in each of the seven subjects. During the five days before substitution with decaffeinated coffee, only one subject reported one

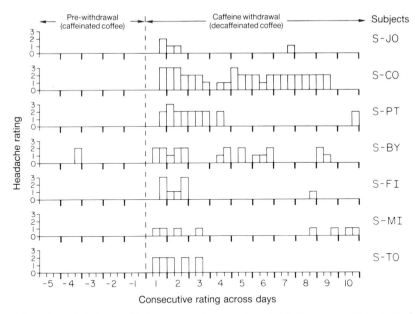

FIG. 9.2. Caffeine withdrawal; effects of substituting decaffeinated for caffeinated coffee on headache rating in each of seven subjects. y-axes: subject-rated headache (0=definitely does not apply; 3=very strongly applies). x-axes: consecutive ratings; ratings were conducted three times each day (8.30 a.m., 12.30 p.m., and 8.30 p.m.); consecutive days are indicated by numerals on x-axes. (Figure reprinted from Griffiths *et al.* 1986*a*.)

instance of headache. On the first day of substitution, one subject reported headache at the 8.30 a.m. rating time, three reported headache at 12.30 p.m., and all seven reported headache at the 8.30 p.m. rating time. The time between the last cup of caffeinated coffee and first report of headache averaged 18.8 ± 1.6 hr (mean ± S.E.) and ranged between 13 and 23 hr across the seven subjects. Although reports of headache were most frequent on the first three days after substitution, there were substantial differences across the subjects with respect to the continuance of headache. For example, in one subject withdrawal headache resolved after three days of decaffeinated coffee while another subject reported almost continuous headache throughout the first nine days after substitution.

Information about the replicability of headache as a caffeine-withdrawal symptom was also provided in three of these same subjects.

After completion of the initial 13 to 17 day decaffeinated substitution phase, caffeinated coffee was reinstated for 11 to 15 days, during which subjects consumed an average of 1.26 g/day caffeine. This was followed by another double-blind substitution of decaffeinated coffee. All three subjects reported a re-emergence of headache on the first one to three days of the second withdrawal period. For all three subjects, however, the frequency of headache was lower during the second withdrawal period than during the first. Although some form of behavioural tolerance or adaptation cannot be ruled out, the finding suggests that two weeks of high-dose caffeine exposure may not be sufficient for establishing the maximal degree of caffeine physical dependence.

REINFORCING EFFECTS IN LABORATORY ANIMALS

Reinforcing efficacy of a drug refers to the relative effectiveness in maintaining behaviour on which the delivery of drug is dependent (Griffiths *et al.* 1979). Over the last 20 years reliable experimental models of drug-taking behaviour in laboratory animals have been developed which provide valid information about the relative reinforcing properties of psychoactive drugs (Griffiths *et al.* 1980).

Six studies of intravenous self-injection in non-human primates (Deneau *et al.* 1969; Griffiths *et al.* 1979; Schuster *et al.* 1969; Yanagita 1970, 1977; Hoffmeister and Wuttke 1973) and five oral and intravenous self-administration studies in rats (Atkinson and Enslen 1976; Collins *et al.* 1984; Heppner *et al.* 1986; Vitiello and Woods 1975; Schulte-Daxboek and Opitz 1981) have not consistently shown caffeine to be self-administered. This literature, which has been reviewed recently (Griffiths and Woodson, in press, *b*), indicates that caffeine does not maintain self-administration behaviour as reliably as classic drugs of abuse such as cocaine, *d*-amphetamine or pentobarbitone. The fact that caffeine does maintain self-administration behaviour under some conditions differentiates caffeine from a wide range of behaviourally-active drugs (including the amphetamine analogue, fenfluramine, and chlorpromazine) which do not maintain self-administration under a variety of conditions (Griffiths *et al.* 1979, 1980).

Figure 9.3 illustrates the results of an intravenous self-injection experiment in baboons (Griffiths *et al.* 1979) which demonstrated the reinforcing effects of caffeine. As shown in the figure, at a dose of 3.2 mg/kg/injection, caffeine maintained steady or erratic daily patterns of self-injection in all three baboons tested; however self-injection was clearly lower and less consistent than during the cocaine control period.

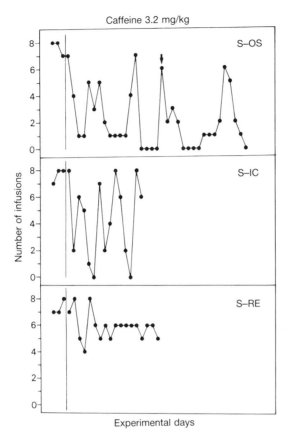

FIG. 9.3. Daily pattern of self-infusion maintained by 3.2 mg/kg/infusion caffeine in three baboons. The initial three-day period in each panel shows the number of infusions maintained by cocaine (0.4 mg/kg/infusion) before substitution of caffeine. The arrow on Day 22 for Baboon S-OS indicates a day on which the animal received one forced infusion of caffeine.
(Figure reprinted from Griffiths *et al.* 1979.)

REINFORCING EFFECTS IN HUMANS

There is a sizeable research literature evaluating various subjective effects of caffeine that might plausibly be related to reinforcing properties. This literature shows that, in contrast to amphetamine (which generally produces prominent elevations in ratings indicating 'euphoria' and 'well-being') caffeine does not consistently produce such effects (Weiss and Laties 1962; Bättig 1985). In fact, a number of studies showed that caffeine produced 'dysphoric' changes in mood, such as increases in anxiety and nervousness (Rapoport *et al.* 1981; Chait and Griffiths 1983; Charney *et al.* 1984).

Goldstein and his colleagues (1969; Goldstein and Kaizer 1969) provided some of the best initial evidence for positive caffeine-induced mood changes in a survey and a clinical pharmacology study. These showed that, after overnight caffeine abstinence, heavy coffee users (five or more cups per day) reported pleasant and desirable effects of coffee drinking and caffeine administration; in contrast, subjects with histories of coffee abstention reported unpleasant and undesirable effects.

The reinforcing properties of caffeine in humans have been investigated more directly by adapting procedures developed in the animal drug self-administration laboratory (Griffiths *et al.* 1980). A recent literature review (Griffiths and Woodson, in press, *b*) indicates that five reports provide information about the behavioural-reinforcing effects of caffeine in humans (Kozlowski 1976; Podboy and Malloy 1977; Griffiths *et al.* 1986*a,b*; Griffiths and Woodson, in press, *b*). These reports suggest that: 1. caffeine is a reinforcer; 2. that the reinforcing effects of caffeine are potentiated by recent histories of heavy caffeine intake (i.e. physical dependence); and 3. there are substantial differences between individual subjects in the reinforcing effects of caffeine (Griffiths and Woodson, in press).

The clearest demonstration of these effects was provided in recent experiments that investigated the self-administration and reinforcing effects of caffeine in coffee in subjects who resided on a research ward (Griffiths *et al.* 1986*a,b*). All subjects reported histories of heavy coffee drinking (mean: 14 cups of coffee per day), and most reported histories suggesting problems with alcohol drinking and/or drug abuse. When cups of coffee were freely available, stable day-to-day patterns of coffee consumption emerged (Fig. 9.4). Coffee drinking tended to be rather regularly spaced during the day, with inter-cup intervals becoming longer throughout the day. Figure 9.5 shows that inter-cup interval progressively increased from approximately 20 min to 80 min over successive cups. Cup duration (i.e., time to complete a cup of coffee), in contrast, showed only a modest increase from the first to the second cup of the day, remaining relatively stable over subsequent cups. Experimental manipulation of coffee concentration, caffeine concentration, and caffeine pre-load showed that this lengthening of inter-cup interval was not due to accumulating caffeine levels. These manipulations also provided evidence for the suppressive effects of caffeine on coffee drinking. Examination of total daily coffee and caffeine intake across manipulations, however, provided no evidence for precise regulation (i.e., titration) of coffee or caffeine intake.

In one experiment, the reinforcing properties of caffeine were

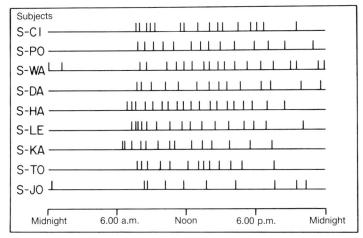

FIG. 9.4. Representative distributions of coffee drinking in each of nine subjects during an *ad-libitum* coffee-drinking phase. Times at which individual cups of coffee were dispensed are represented by vertical hatch marks.
(Figure reprinted from Griffiths *et al.* 1986*b*.)

evaluated by substituting decaffeinated coffee for caffeinated coffee (100 mg/cup) under double-blind conditions (Griffiths *et al.* 1986*a*). It was reasoned that, if coffee drinking were maintained predominantly by caffeine, then perhaps coffee drinking would progressively decrease (due to behavioural extinction) with prolonged exposure to decaffeinated coffee. Three subjects were exposed to phases involving the sequential availability of caffeinated, decaffeinated, caffeinated, decaffeinated, and caffeinated coffee, with decaffeinated phases being up to 13 to 17 days in each subject. An additional four subjects were exposed to phases of ten or more days of decaffeinated coffee after a period of continuous exposure to caffeinated coffee. These double-blind manipulations provided no evidence for the behavioural extinction of coffee drinking in the decaffeinated coffee phases. On the first few days after substitution of decaffeinated coffee, coffee-drinking decreased slightly (but non-significantly), while coffee-liking decreased significantly on the first two days after substitution, and subsequently progressively increased to pre-substitution levels. This transient decrease in liking was probably due to caffeine withdrawal, which was measured concurrently on other subjective and objective scales and showed a similar time-course. As a whole, the experiment provided no evidence for progressive lessening of coffee drinking, as' would be predicted on the basis of behavioural extinction. To the extent that liking might predict reinforcing effects, the experiment did provide suggestive evidence that, relative to caffeinated

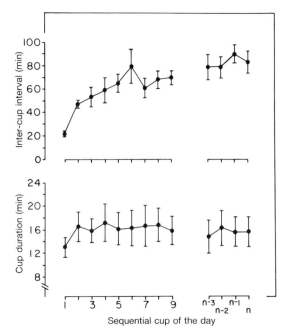

FIG. 9.5. Average inter-cup interval (top) and cup duration (bottom) as a function of sequential cup of the day. y-axes: inter-cup interval and cup duration in minutes. x-axes: sequential cup of the day. The first cup to be drunk after 6.00 a.m. was considered the first cup of the day; n indicates the last cup of the day. Data points and brackets indicate means ±1 S.E.M. for nine subjects during a six-day *ad-libitum* coffee-drinking phase.
(Figure reprinted from Griffiths *et al.* 1986*b*.)

coffee, decaffeinated coffee may be aversive in subjects physically dependent on caffeine.

Since the results of the preceding experiment showed that a free self-administration approach was insensitive to possible differences in the reinforcing properties of caffeine in coffee, another experiment was undertaken which utilized a choice procedure to explore the relative reinforcing effects of caffeinated (100 mg per cup) versus decaffeinated coffee (Griffiths *et al.* 1986*a*). On some days ('no-choice' days) the available coffee was identified to subjects and staff by a letter code. On 'choice' days, two letter-coded coffees (to which the subject had previously been exposed) were available for consumption; subjects made a mutually-exclusive choice as to which lettered coffee would be consumed that day. Subjects were exposed to these choice tests under two different 'background' conditions (i.e., double-blind caffeinated or decaffeinated background condition in which subjects consumed only caffeinated or decaffeinated coffee for a week or more before the first choice test).

Of the twelve choice tests (in six subjects) conducted under the caffeinated background condition, caffeinated coffee was overwhelmingly preferred (92 per cent) to decaffeinated coffee. In contrast, caffeinated coffee was not reliably chosen under the decaffeinated background condition. Of the eight choice tests (in four subjects) conducted under the decaffeinated background condition, caffeinated and decaffeinated coffee were chosen equally often (50 per cent). Of the four subjects examined, one chose only caffeinated coffee, two chose only decaffeinated coffee, and one chose both caffeinated and decaffeinated coffee on different occasions.

Figure 9.6 shows that the liking ratings from the 'no-choice' days were consistent with the behavioural choice results. Under the caffeinated background condition, caffeinated coffee was rated as better liked than the decaffeinated coffee, which was rated very unfavourably. All six subjects complained that the decaffeinated coffee had low strength or

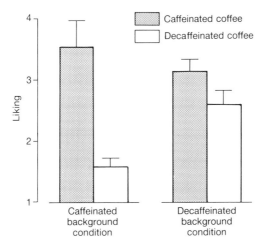

FIG. 9.6. Subject-rated liking of caffeinated coffee (filled bars) and de-caffeinated coffee (open bars) on 'no-choice' days that preceded choice opportunities in the choice sequences. y-axes: 8.30 p.m. ratings of liking. Each bar shows mean, and each bracket shows 1 S.E.M. for N=6 subjects under the caffeinated background condition, and N=4 under the decaffeinated background condition. For purposes of data presentation, when a subject received more than one exposure to a choice sequence, mean subject data was used.
(Figure reprinted from Griffiths and Woodson, in press, *b.*)

stimulant effect, and several subjects attributed dysphoric symptoms (e.g., fatigue and headache, which were probably due to caffeine withdrawal) to the decaffeinated coffee. Under the decaffeinated background condition, however, there was no such pronounced difference in liking between caffeinated and decaffeinated coffee. Subjects in this condition indicated that the decaffeinated coffee was satisfactory, and they did not complain about the lack of strength or stimulant effect.

Comparing across the two background conditions in Figure 9.6, it appears that the difference in liking under the caffeinated background condition is due primarily to a decreased liking for the decaffeinated coffee rather than a change in liking for the caffeinated coffee. This result is consistent with the transient decrease in liking that occurred in the preceding experiment, and is probably attributable to caffeine withdrawal.

CONCLUSION

Caffeine is the most widely-used psychotropic agent in the world. Given the high prevalence and long, historical persistence of caffeine use, it is not surprising that it has periodically been identified as a drug of abuse. As discussed earlier, drugs of abuse have two major characteristics;

(1) they produce adverse effects (they have the capacity to harm the individual and/or society), and

(2) they have reinforcing properties (Griffiths *et al.* 1985).

The present review documents both types of characteristics with caffeine. Caffeine has been shown to produce substantial adverse disruptions in mood and behaviour upon its withdrawal, and to serve as a reinforcer.

Physical dependence is among the most clearly-demonstrated adverse effects of habitual caffeine use. Such dependence is manifested by biochemical, physiological, behavioural, and subjective disturbances which occur upon termination of chronic drug administration. In the six published reports of caffeine withdrawal in laboratory animals, the only behavioural effect to be reliably documented is decreased locomotor activity in rodents. In contrast, human studies have clearly and repeatedly documented a caffeine withdrawal syndrome. From the thirty-seven reports published to date, this syndrome may be characterized predominantly by the occurrence of headache and fatigue, with a wide variety of other signs and symptoms occurring at a lower frequency.

With respect to the behaviour-reinforcing effects of caffeine, there have been eleven pre-clinical studies published to date. Although caffeine does not maintain self-administration behaviour as reliably as classic drugs of abuse, the fact that caffeine does maintain self-administration behaviour under some conditions differentiates it from a wide range of behaviourally-active compounds (e.g. fenfluramine and chlorpromazine) which do not maintain such behaviour under a variety of conditions. With respect to human research, five caffeine self-administration studies and several studies evaluating subjective effects have shown that, under appropriate conditions, caffeine can serve as a reinforcer, and can produce elevations in subjective drug-liking and/or euphoria. A large number of studies evaluating the subjective effects of caffeine, however, suggest that caffeine is not as effective as amphetamine in producing increases in euphoria or well-being. In fact, a number of studies showed that caffeine produced dysphoric changes in mood, such as increases in anxiety and one self-administration study showed that there may be substantial differences between individual subjects in the reinforcing effect of caffeine.

Several human studies and one animal experiment suggest that physical dependence may substantially potentiate the reinforcing effects of caffeine.

In addition to a withdrawal syndrome, the extent of additional health risk associated with societally-sanctioned levels of caffeine use is controversial (Gilbert 1976; Greden 1981; Dews 1982; Ernster 1984). Since the relative abuse liability of a drug can be considered to be a multiplicative function of the degree of reinforcing properties and adverse effects (Griffiths *et al.* 1985), a meaningful classification of caffeine, relative to more widely-recognized drugs of abuse, should await resolution of the health risk controversy. Until that time, however, it would be prudent if there were greater societal recognition that caffeine has the cardinal features of a prototypic drug of abuse.

ACKNOWLEDGEMENT

This research was supported in part by U.S.P.H.S. research grants RO1 DA03890 and RO1 DA01147 from the National Institute on Drug Abuse. Sections of this chapter were adapted from Griffiths and Woodson (in press, *a,b*). Send reprint requests to Roland Griffiths, Psychiatry, The Johns Hopkins University School of Medicine, 720 Rutland Ave, Baltimore, MD 21205.

REFERENCES

Atkinson, J. and Enslen, M. (1976). Self-administration of caffeine by the rat. *Arzneimittel-Forschung/Drug Research* **26**, 2059–61.

Austin, G. A. (1979). *Perspectives on the history of psychoactive substance use.* Research Issues **24**, DHEW Publication No. (ADM) 79-810, pp. 50–66. US Government Printing Office, Washington DC.

Bättig, K. (1985). The physiological effects of coffee consumption. In *Coffee: Botany, biochemistry and production of beans and beverage* (eds. M. N. Clifford and K. C. Willson), pp. 394–439. The AVI Publishing Company, Inc., Westport, Connecticut.

Boyd, E. M., Dolman, M., Knight, L. M., and Sheppard, E. P. (1965). The chronic oral toxicity of caffeine. *Canadian J. Physiol. and Pharmacol.* **43**, 995–1007.

Brady, J. V. and Lukas, S. E. (eds.) (1984). *Testing drugs for physical dependence potential and abuse liability.* National Institute on Drug Abuse Research Monograph, No. 52. DHHS Publication No. (ADM) 84-1332. US Government Printing Office, Washington DC.

Carney, J. M. (1982). Effects of caffeine, theophylline and theobromine on scheduled controlled responding in rats. *British J. Pharmacol.* **75**, 451–4.

Chait, L. D. and Griffiths, R. R. (1983). Effects of caffeine on cigarette smoking and subjective response. *Clin. Pharmacol. and Ther.* **34**, 612–22.

Charney, D. S., Galloway, M. P., and Heninger, G. R. (1984). The effects of caffeine on plasma MHPG, subjective anxiety, autonomic symptoms and blood pressure in healthy humans. *Life Sciences* **35**, 135–44.

Cobbs, L. W. (1982). Lethargy, anxiety, and impotence in a diabetic. *Hospital Practice* **17**(8), 67, 70, 73.

Collins, R. J., Weeks, J. R., Cooper, M. M., Good, P. I., and Russell, R. R. (1984). Prediction of abuse liability of drugs using IV self-administration by rats. *Psychopharmacology* **82**, 6–13.

Deneau, G., Yanagita, T., and Seevers, M. H. (1969). Self-administration of psychoactive substances by the monkey: A measure of psychological dependence. *Psychopharmacologia (Berlin)* **16**, 30–48.

Dews, P. B. (1982). Caffeine. *Ann. Review of Nutrition* **2**, 323–41.

Dreisbach, R. H. and Pfeiffer, C. (1943). Caffeine-withdrawal headache. *J. Laboratory and Clinical Medicine* **28**, 1212–19.

Ernster, V. L. (1984). Epidemiologic studies of caffeine and human health. In *The methylxanthine beverages and foods: Chemistry, consumption, and health effects* (ed. G. A. Spiller), pp. 377–400. Alan R. Liss, New York.

Finn, I. B. and Holtzman, S. G. (1986). Tolerance to caffeine-induced stimulation of locomotor activity in rats. *J. Pharmacol. and Exper. Ther.* **238**, 542–6.

Gilbert, R. M. (1976). Caffeine as a drug of abuse. In *Research advances in alcohol and drug problems*, Vol. 3. (eds. R. J. Gibbins, Y. Israel, H. Kalant, R. E. Popham, W. Schmidt, and R. G. Smart), pp. 49–176. John Wiley and Sons, New York.

—— (1984). Caffeine consumption. In *The methylxanthine beverages and foods: Chemistry, consumption, and health effects* (ed. G. A. Spiller), pp. 185–213. Alan R. Liss, New York.

Goldstein, A. and Kaizer, S. (1969). Psychotropic effects of caffeine in man. III. A questionnaire survey of coffee drinking and its effects in a group of housewives. *Clin. Pharmacol. and Ther.* **10**, 477–88.

——, ——, and Whitby, O. (1969). Psychotropic effects of caffeine in man. IV. Quantitative and qualitative differences associated with habituation to coffee. *Clin. Pharmacol. and Ther.* **10**, 489–97.

Graham, D. M. (1978). Caffeine—Its identity, dietary sources, intake and biological effects. *Nutrition Reviews* **36**, 97–102.

Graham, H. N. (1984). Tea: The plant and its manufacture; chemistry and consumption of the beverage. In *The methylxanthine beverages and foods: Chemistry, consumption, and health effects* (ed. G. A. Spiller), pp. 29–74. Alan R. Liss, New York.

Greden, J. F. (1981). Caffeinism and caffeine withdrawal. In *Substance abuse: Clinical problems and perspectives* (eds. J. H. Lowinson and P. Ruiz), pp. 274–86. Williams and Wilkins, Baltimore.

——, Victor, B. S., Fontaine, P., and Lubetsky, M. (1980). Caffeine-withdrawal headache: A clinical profile. *Psychosomatics* **21**, 411–13; 417–18.

Griffiths, R. R. and Woodson, P. P. (in press, *a*). Caffeine physical dependence: Studies in humans and laboratory animals. *Psychopharmacology.*

—— and —— (in press, *b*). Reinforcing properties of caffeine: Studies in humans and laboratory animals. *Pharmacology, Biochemistry and Behavior.*

——, Brady, J. V., and Bradford, L. D. (1979). Predicting the abuse liability of drugs with animal drug self-administration procedures: Psychomotor stimulants and hallucinogens. In *Advances in behavioral pharmacology*, Vol. 2. (eds. T. Thompson and P. B. Dews), pp. 163–208. Academic Press, New York.

——, Bigelow, G. E., and Henningfield, J. E. (1980). Similarities in animal and human drug-taking behavior. In *Advances in substance abuse*, Vol. 1. (ed. N. K. Mello), pp. 1–90. JAI Press, Greenwich, Connecticut.

——, Lamb, R. J., Ator, N. A., Roache, J. D., and Brady, J. V. (1985). Relative abuse liability of triazolam: Experimental assessment in animals and humans. *Neuroscience and Biobehavioral Reviews* **9**, 133–51.

——, Bigelow, G. E., and Liebson, I. A. (1986*a*). Human coffee drinking: Reinforcing and physical dependence producing effects of caffeine. *J. Pharmacol. and Exper. Ther.* **239**, 416–25.

——, ——, ——, O'Keeffe, M., O'Leary, D., and Russ, N. (1986*b*). Human coffee drinking: Manipulation of concentration and caffeine dose. *J. Exper. Anal. of Behavior* **45**, 133–48.

Heppner, C. C., Kemble, E. D., and Cox, W. M. (1986). Effects of food deprivation on caffeine consumption in male and female rats. *Pharmacology, Biochemistry and Behavior* **24**, 1555–9.

Hoffmeister, F. and Wuttke, W. (1973). Self-administration of acetylsalicylic acid and combinations with codeine and caffeine in rhesus monkeys. *J. Pharmacol. and Exper. Ther.* **186**, 266–75.

Holtzman, S. G. (1983). Complete, reversible, drug-specific tolerance to stimulation of locomotor activity by caffeine. *Life Sciences* **33**, 779–87.

—— and Finn, I. B. (in press). Tolerance to behavioral effects of caffeine in rats. *Pharmacology, Biochemistry and Behavior.*

Kozlowski, L. T. (1976). Effect of caffeine on coffee drinking. *Nature* **264**, 354–5.

Martin, W. R. (ed.) (1977). *Drug addiction I: Morphine, sedative/hypnotic and alcohol dependence, Handbook of experimental pharmacology*, **45/I**. Springer-Verlag, Berlin.

Petursson, H. and Lader, M. (1984). *Dependence on tranquillizers*. Institute of Psychiatry Maudsley Monographs, No. 28. Oxford University Press, Oxford.

Podboy, J. and Malloy, W. (1977). Caffeine reduction and behavior changes in the severely retarded. *Mental Retardation* **15**, 40.

Rainey, J. T. (1985). Headache related to chronic caffeine addiction. *Texas Dental Journal* **102**, 29–30.

Rapoport, J. L., Jensvold, M., Elkins, R., Buchsbaum, M. S., Weingartner, H., Ludlow, C., Zahn, T. P., Berg, C. J., and Neims, A. H. (1981). Behavioral and cognitive effects of caffeine in boys and adult males. *J. Nervous and Mental Disease* **169**, 726–32.

Schulte-Daxboek, G. and Opitz, K. (1981). Increased caffeine consumption following chronic nicotine treatment in rats. *IRCS Medical Science* **9**, 1062.

Schuster, C. R., Woods, J. H., and Seevers, M. H. (1969). Self-administration of control stimulants by the monkey. In *Abuse of central stimulants* (eds. F. Sjoqvist and M. Tottie), pp. 339–47. Raven Press, New York.

Vitiello, M. V. and Woods, S. C. (1975). Caffeine: Preferential consumption by rats. *Pharmacology, Biochemistry and Behavior* **3**, 147–9.

—— and —— (1977). Evidence for withdrawal from caffeine by rats. *Pharmacology, Biochemistry and Behavior* **6**, 553–5.

Weiss, B. and Laties, V. G. (1962). Enhancement of human performance by caffeine and the amphetamines. *Pharmacological Reviews* **14**, 1–36.

Wilkin, J. K. (1986). The caffeine withdrawal flush: Report of a case of "weekend flushing". *Military Medicine* **151**, 123–4.

Yanagita, T. (1970). Self-administration studies on various dependence-producing agents in monkeys. *University of Michigan Medical Center J.* **36**, 216–24.

—— (1977). Brief review on the use of self-administration techniques for predicting drug dependence potential. In *Predicting dependence liability of stimulant and depressant drugs* (eds. T. Thompson and K. R. Unna), pp. 231–42. University Park Press, Baltimore.

10

The essence of benzodiazepine dependence

SIOBHAN M. MURPHY and PETER TYRER

INTRODUCTION

In recent years there have been many reports in the medical literature suggestive of addiction to benzodiazepines. Although various words are used to describe these problems, including 'dependence' (Ashton 1984; Rickels *et al.* 1986), 'rebound anxiety' (Power *et al.* 1985; Fontaine *et al.* 1984), 'rebound insomnia' (Kales *et al.* 1978, 1979), and 'withdrawal or abstinence symptoms' (Power *et al.* 1985), almost all these describe phenomena that occur only after withdrawal of the drug. By contrast there is a relatively small number of reports of patients taking these drugs in excessive dosage, accompanied by gross drug-seeking behaviour and rapid development of tolerance. The relative rarity of such reports had led Marks in the late 1970s to conclude that significant dependence on benzodiazepines was not a major clinical problem (Marks 1978).

Most recent reports of dependence have been concerned with the use of these drugs in therapeutic dosage. The 1970s could well be described as the 'tranquillizer era', as benzodiazepines were prescribed more commonly than any other drugs (Skegg *et al.* 1977). However, the demonstration of dependence in normal dosage has created a tidal wave of criticism, and calumny is being heaped on those who described these as wonder drugs when they were introduced in the 1960s. Benzodiazepines are now tainted in the eyes of the general public and medical profession alike; instead of being perceived as safe and harmless, they have the image of being grossly addictive. A complete ban on their prescription has even been suggested (Snaith 1984). This is a fate which has never been suggested for antidepressants, antipsychotic drugs, lithium, or antihypertensive agents, despite good evidence that at least some symptoms of withdrawal can occur when they are stopped suddenly (Kramer *et al.* 1961; Gardos *et al.* 1978; Klein *et al.* 1981; Cole *et al.* 1979). However, this criticism is not universal; at the opposite pole, it has been suggested recently that it would be more appropriate to have these drugs available to the general public without prescription (Oswald 1986). It therefore seems to be an appropriate time to examine what exactly constitutes benzodiazepine dependence, and how it can be recognized and avoided.

WHAT IS DEPENDENCE?

It is difficult to decide whether an individual or a group is dependent on benzodiazepines unless one is clear exactly what is meant by the term 'dependence'. Unfortunately this remains a somewhat controversial issue (Edwards *et al.* 1982). The definition of dependence by the World Health Organization International Classification of Diseases (1977) as, 'characterised by behavioural and other responses that always include a compulsion to take the drug on a continuous or periodic basis in order to experience its psychic effects, and sometimes to avoid the discomfort of its absence', is admirable in many respects. However, it stresses the psychic effects of the drug and, bearing in mind that all psychotropic drugs are given for their effects on the psyche, it is sometimes difficult to decide whether the 'compulsion to take the drug' is a consequence of addiction or a therapeutic need.

Russell (1976*b*) regards the presence of physical dependence, and the presence of an abstinence syndrome after withdrawal, as crucial to both the definition and severity of the dependent state. Thus dependence 'requires the crucial feature of negative affect experienced in its absence. The degree of dependence can be equated with the amount of negative affect, which may range from mild discomfort to extreme distress, which may be equated with the amount of difficulty or effort required to do without the drug'. This would certainly be a good description of many patients who are dependent on benzodiazepines.

The interactional components of drug, person, and environment have been emphasized by Edwards and his colleagues (1982) in their definition of dependence as a 'psycho-physiological-social' syndrome including;

(1) subjective awareness of compulsion to use a drug, usually during attempts to stop or moderate drug use;

(2) a desire to stop drug use in the face of continued use;

(3) a relatively stereotyped drug-taking habit;

(4) evidence of neuro-adaptation (tolerance and withdrawal effects);

(5) the salience of drug-seeking behaviour relative to other important priorities, and

(6) rapid reinstatement of the syndrome after a period of abstinence.

Descriptions of physical and psychological dependence are common to all these definitions and, in most of them, tolerance is also a factor. Psychological dependence emphasizes the pleasurable reinforcing effects of the drug and avoidance of discomfort by its reduction or withdrawal, whereas physical dependence emphasizes the intense physical disturbances that occur when the administration of the drug is suspended (Eddy *et al.* 1965). The withdrawal syndrome that accompanies physical dependence can be described as 'a well-defined syndrome with predictable onset, duration, and offset of action containing psychological and bodily symptoms not previously complained of by the patient. This can be repressed by reinstituting the drug, and is drug specific' (Lader 1983).

Evidence of dependence with benzodiazepines

For the purposes of this chapter most of the discussion is limited to dependence on benzodiazepines in therapeutic dosage. There is no doubt that benzodiazepines can be abused, particularly by polydrug users, but quantitatively this is rare compared with the vast numbers of patients who are taking benzodiazepines in therapeutic dosage. How many of these are dependent, and how can it be decided whether their drugs are fulfilling a therapeutic need or merely maintaining the dependent cycle?

Psychological dependence
The two halves of psychological dependence are the induction of pleasure by the drug in question and the avoidance of discomfort by its continued use. Although the consumption of benzodiazepines is sometimes followed by clear drug-seeking behaviour (Hanna 1972; Winstead *et al.* 1974), it is usually a background feature and, by comparison with other addictive drugs, is unimportant. Drug-seeking behaviour is associated with a history of abuse with other drugs and alcohol (Marks 1978). However, there is no doubt that large numbers of patients take benzodiazepines in therapeutic dosage regularly to avoid discomfort in the broader sense. Whether such discomfort constitutes a return to the anxious state existing before the drugs were ever prescribed, or whether it is the phase of a withdrawal syndrome, is rarely known. Many patients prefer not to put this to the test and therefore go on taking the drugs in regular dosage whenever these mild symptoms of discomfort appear.

In animal studies using self-administration designs some mammals, particularly monkeys, demonstrate increased consumption of benzodiazepines compared with placebo, but this is somewhat less than that with barbiturates (Harris *et al.* 1968; Findley *et al.* 1972; Yanagita 1982). Even in these experiments, it is difficult to know whether self-administration is carried out for purposes of inducing pleasure or for avoidance of unpleasant symptoms on withdrawal.

Tolerance and neuro-adaptation

Tolerance can be pharmacodynamic or pharmacokinetic. There is no evidence that benzodiazepines either induce or inhibit their own meta-bolism (Greenblatt *et al.* 1979*a*), nor do they affect the rate of drug metabolism in the liver (Vessel and Penno 1983); after prolonged use, plasma concentrations remain constant (Greenblatt *et al.* 1981*b*; Rutherford *et al.* 1978).

However, there is a large amount of evidence that shows that benzo-diazepines induce pharmacodynamic tolerance (Rosenberg and Chiu 1985). This is most obvious after a large overdose of benzodiazepines; consciousness returns at plasma concentrations that would normally be associated with coma (Greenblatt *et al.* 1979*b*, 1981*a*). Tolerance also develops to the electroencephalographic effects of benzodiazepines (Lader *et al.* 1980) and, after chronic use, the growth-hormone response to benzodiazepine challenge is suppressed (Shur *et al.* 1983). However, with most patients, tolerance does not continue to develop and so patients can take the same dosage for many years, once the initial tolerance to the drug has developed.

The exact nature of this tolerance is difficult to determine. The most obvious explanation would be a change in the sensitivity of benzo-diazepine receptors. The presence of these receptors, discovered over ten years ago (Squires and Braestrup 1977; Squires *et al.* 1979), with their high concentrations in the limbic system, cerebral cortex, and cerebellum (Möhler and Okada 1977) suggest that there is an endogenous benzodiazepine-like substance (or antagonist) that is responsible for normal physiological control of anxiety. However this endogenous ligand has yet to be found, if it indeed exists. Benzodiazepines facilitate the important inhibitory neurotransmitter, gamma-aminobutyric acid (GABA) through its close affinity with GABA receptors and the chloride channel in an ionophore (Braestrup and Neilsen 1980). Stimulation of GABA pathways decreases neurotransmission in acetylcholine, dopa-mine, and serotonergic pathways (Iversen 1978; Collinge *et al.* 1982), but the selectivity of the benzodiazepine–GABA complexes ensures that generalized central depression does not take place. There have also been suggestions that at least some of the action of benzodiazepines could be mediated through endogenous opioid peptides (Millan and Duka 1981).

Although benzodiazepine receptors have been studied in depth for over ten years, we are still no further towards explaining the phenomenon of tolerance. Although there are clues that a benzodiaze-pine receptor antagonist could be an endogenous ligand for these receptors (Woolf and Nixon 1981) there is still no evidence that these substances bind directly with GABA-benzodiazepine receptor com-plexes. The most obvious explanations of tolerance would be an increase in the number of benzodiazepine receptors, or a decrease in their sensitivity (Braestrup *et al.* 1979; Shur *et al.* 1983). It has also been suggested (Ashton 1984) that tolerance could be caused by down-regulation of GABA receptors; this seems to be a more likely explanation than one involving the benzodiazepine receptors primarily. It has been demonstrated that GABA-receptor density is decreased, and that a syndrome very similar to that of benzodiazepine withdrawal can be precipitated by the administration of GABA antagonists (Crawley *et al.* 1982; Cowen and Nutt 1982). Indeed it has been postulated that all withdrawal syndromes associated with addiction to antianxiety drugs (including alcohol, barbiturates, and benzodiazepines) could be explained by a common GABA mechanism (Cowen and Nutt 1982).

The major difficulty in defining the benzodiazepine withdrawal syndrome is that of separating the symptoms of withdrawal from those of the anxiety that may have existed before the drug was ever prescribed. There are three possible consequences to stopping a tranquillizing drug; no change in symptoms, a return to previous levels of anxiety (which is then maintained), or a withdrawal reaction (Owen and Tyrer 1983). The distinction between the latter two can only be made if treatment is withheld and an abstinence syndrome develops, with symptoms that are qualitatively different from those clinical anxiety. The combination of increasing anxiety plus novel withdrawal symptoms is now used most commonly in recent reports that define withdrawal (Rickels *et al.* 1986; Power *et al.* 1985; Fontaine *et al.* 1984; Tyrer *et al.* 1981, 1983; Busto *et al.* 1986), whilst this distinction was often not made in earlier reports (Petursson and Lader 1981; Winokur *et al.* 1980).

As might be expected, withdrawal symptoms are generally more severe when high, rather than low, dosage has been taken. The most severe, and unequivocal, withdrawal symptoms are those of epileptic seizures, psychotic symptoms (in which paranoid features predominate) and delirium (Hollister *et al.* 1961; Preskorn and Denner 1977). These can occur after stopping benzodiazepines in therapeutic dosage, but are less common. After lower dosage, the most common withdrawal symptoms are those of increased anxiety, insomnia, irritability, nausea, palpitations, tremor, muscular tension, headache, and dysphoria. Additional symptoms include tinnitus, parasthaesia, hypersensitivity to sensory stimuli, muscle twitching, perceptual abnormalities, depersonalization, and derealization. It is this 'middle ground' of symptoms that causes the most argument over whether they constitute a withdrawal reaction. None of these symptoms can be regarded as pathognomonic of a withdrawal syndrome but some symptoms, particularly the perceptual ones, appear to be much more prominent in benzodiazepine withdrawal than in normal anxiety syndromes.

The period of onset of the withdrawal reactions varies between three and seven days, and this appears to depend mainly on the elimination half-life of the benzodiazepine (Busto and Sellers 1986). The major symptoms of withdrawal last between five and 20 days, after which there is gradual resolution. However, there is a period of residual anxiety and decreased capacity to cope with stressful situations for at least several weeks, and possibly up to six months, in patients who have had withdrawal reactions (Ashton 1984; Owen and Tyrer 1983). In behavioural terms, this could be explained as a period of learning to cope with stress, after a prolonged period in which the coping has been carried out by an exogenous drug.

Predicting dependence with benzodiazepines

Unfortunately, there are no firm predictors of dependence, although, as mentioned earlier, there are several risk factors which make dependence more likely. These include dosage, with increased likelihood of dependence above the therapeutic range (Ayd 1979) but, at the other end of the spectrum, there does not appear to be a completely safe dosage in which a benzodiazepine can be taken regularly without any risk of dependence. Regular, rather than intermittent, dosage increases the chance of developing dependence (Tyrer and Seivewright 1984) and, in general, the longer the drug has been taken the more likely is dependence to be a problem (Ayd 1979). However, evidence of withdrawal symptoms have been reported after regular dosage between four weeks and 16 years, and rebound insomnia has been reported after only a few nights of medication with a benzodiazepine hypnotic (Kales *et al.* 1978).

There is also some evidence that personality characteristics are related to the development of dependence (Rickels *et al.* 1986; Tyrer *et al.* 1983; Rickels *et al.* 1984). Studies using the Minnesota Multiphasic Personality Inventory (MMPI) and the Personality Assessment Schedule (PAS) have shown that traits and behaviour linked to dependence and resourceless-ness are more common in patients who have withdrawal symptoms. In patients with normal MMPI profiles, no withdrawal symptoms were experienced after stopping benzodiazepines (Rickels *et al.* 1984) and this may explain why patients taking benzodiazepines for medical conditions (such as spasticity and epilepsy) are less prone to withdrawal reactions.

Although there is a large range of figures quoted in the literature, it appears that most unselected patients taking benzodiazepines primarily for anxiety and insomnia would include between 30 and 40 per cent who would have a withdrawal reaction when they reduce their medication (Tyrer *et al.* 1981).

COMPARISON OF BENZODIAZEPINE DEPENDENCE WITH OTHER FORMS OF DRUG ABUSE

Benzodiazepine dependence is characterized by early development of tolerance with little subsequent change, little or no escalation of dosage, and a withdrawal syndrome after stopping treatment. Drug-seeking behaviour is present, but not to a marked degree. Comparison with other common forms of drug dependence suggest that benzodiazepines come close to nicotine dependence, as these characteristics are all shared by tobacco users. It is perhaps no coincidence that the difficulties subjects have when stopping nicotine are approximately the same order as those stopping benzodiazepines after prolonged usage, and that there is a similar period of 'post-withdrawal vulnerability' with nicotine that needs to be overcome before the habit is broken entirely (Russell 1976*a*).

Although the most unequivocal evidence of a withdrawal syndrome is the presence of completely new symptoms occurring at the time of withdrawal from the drug, it is very difficult to measure this in practice. If, for whatever reason, subjects are led to believe that they are dependent on a particular drug, they are likely to interpret many of the symptoms following withdrawal as novel and explained only by drug addiction. These symptoms can occur if subjects are led to believe that drug withdrawal is taking place, even when it is not, as for example in the context of double blind studies. In one recent study, 20 per cent of patients experienced significant withdrawal symptoms (using the above definitions) at a time when drug therapy was unchanged (Tyrer *et al.* 1983).

If symptoms are recorded respectively for some weeks before withdrawal takes place, it is possible to detect completely novel symptoms in a more accurate way. For example, in one recent study (Murphy *et al.* 1984), it was found that patients who stopped diazepam after six weeks of regular treatment had a significant increase in anxiety symptoms. Close analysis revealed that almost all this was explained by an increase in pre-existing symptoms of anxiety rather than the expression of novel symptoms. The identification of a qualitatively distinct and pathognomonic withdrawal syndrome from benzodiazepines continues to elude researchers, and may not exist. As clinical measures of withdrawal syndromes are far from satisfactory, further research in this field will be hampered until there are more acceptable indices of dependence that are not affected so much by psychological factors.

CONCLUSION

Benzodiazepine dependence is difficult to identify because most dependent patients only show evidence of addictive behaviour after their drugs are stopped. The demonstration of an unequivocal withdrawal syndrome (consisting of an increase in existing anxious symptoms, the onset of new and sometimes severe symptoms of withdrawal, and the subsequent resolution of these symptoms) is indicative of true dependence. Unfortunately interpretation of these phenomena is hindered as most patients take benzodiazepines for anxious symptoms which may recur naturally after stopping the drug, and because such patients are prone to psychological forms of dependence and pseudowithdrawal reactions. However, there is now good evidence that regular high dosage for longer than six weeks, past history of dependence on other drugs, and dependent-personality characteristics are all associated with greater risk of dependence.

In pharmacological terms, the withdrawal reaction is probably a GABA-deficiency syndrome, and can be simulated by the administration of GABA-antagonists or benzodiazepine antagonists to normal people.

REFERENCES

Ashton, H. (1984). Benzodiazepine withdrawal: an unfinished story. *Brit. Medical J.* **288**, 1135–40.

Ayd, F. J. Jr. (1979). Benzodiazepines: dependence and withdrawal. *J. Amer. Med. Assoc.* **242**, 1401–2.

Braestrup, C. and Neilsen, M. (1980). Benzodiazepine receptors. *Arzneimittel-Forschung* **30**, 853–9.

——, ——, and Squires, R. F. (1979). No changes in rat benzodiazepine receptors after withdrawal from continuous treatment with lorazepam and diazepam. *Life Sciences* **24**, 347–50.

Busto, U. and Sellers, E. (1986). Pharmacokinetic determinants of drug abuse and dependence. A conceptual perspective. *Clinical Pharmacokinetics* **11**, 144–53.

——, ——, Naranjo, C. A., Cappell, H., Sanchez-Craig, M., and Sykora, K. (1986). Withdrawal reaction after long-term therapeutic use of benzodiazepines. *New England J. Medicine* **315**, 854–9.

Cole, J. O., Altesman, R. I., and Weingarten, C. H. (1979). Psychopharmacology update: beta blocking drugs in psychiatry. *McLean Hospital Journal* **4**, 40–67.

Collinge, J., Pycock, C. J. and Taberner, P. V. (1982). GABA- and diazepam-induced reduction of cerebral 5-hydroxytryptamine turnover. *Brit. J. Pharmacol.* **75**, 45.

Cowen, P. J. and Nutt, D. J. (1982). Abstinence symptoms after withdrawal from tranquillising drugs; is there a common neurochemical pathway? *Lancet* **ii**, 360–2.

Crawley, J. N., Marangos, P. J., Stivers, J., and Goodwin, F. K. (1982). Chronic clonazepam administration induces benzodiazepine receptor subsensitivity. *Neuropharmacology* **21**, 85–9.

Eddy, N. B., Halbach, H., Isbell, H., and Seevers, M. A. (1965). Drug dependence, its significance and characteristics. *Bull. World Health Organisation* **32**, 721–33.

Edwards, G., Arif, A., and Hodgson, R. (1982). Nomenclature and classification of drug and alcohol related problems: A shortened version of a World Health Organisation memorandum. *Brit. J. Addiction* **77**, 3–20.

Findley, J. D., Robinson, W. W., and Peregrino, L. (1972). Addiction to secobarbital and chlordiazepoxide in the rhesus monkey by means of a self-infusion preference procedure. *Psychopharmacologia* **26**, 93–114.

Fontaine, R., Chouinard, G., and Annable, L. (1984). Rebound anxiety in anxious patients after abrupt withdrawal of benzodiazepine treatment. *Amer. J. Psychiatry* **141**, 848–52.

Gardos, G., Cole, J., and Tarsy, D. (1978). Withdrawal syndromes associated with antipsychotic drugs. *Amer. J. Psychiatry* **135**, 1321–4.

Greenblatt, D. J., Allen, M. D., MacLaughlin, D. S., Haffman, D. H., Harmatz, J. S., and Schader, R. I. (1979*a*). Single and multiple dose kinetics of oral lorazepam in humans. The predictability of accumulation. *J. Pharmacokinetics and Biopharmaceutics* **7**, 159–79.

——, Schader, R. I., Harmatz, J. S., and Georgotas, A. (1979*b*). Self rated sedation and plasma concentrations of desmethyldiazepam follow single doses of clorazepate. *Psychopharmacology* **66**, 289–90.

——, Divoll, M., Harmatz, J. S., McLaughlin, D. S., and Schader, R. I., (1981*a*). Kinetics and clinical effects of flurazepam in young and elderly non insomniacs. *Clin. Pharmacol. and Ther.* **30**, 475–86.

——, Laughren, T. P., Allen, M. D., Harmatz, J. S., and Schader, R. I. (1981*b*). Plasma diazepam and desmethyldiazepam concentrations during long-term diazepam therapy. *Brit. J. Clin. Pharmacol.* **11**, 35–40.

Hanna, S. M. (1972). A case of oxazepam (Serenid D) dependence. *Brit. J. Psychiatry* **120**, 443–5.

Harris, R. T., Glaghorn, J. L., and Schoolar, J. C. (1968). Self administration of minor tranquillisers as a function of conditioning. *Psychopharmacologia* **13**, 81–8.

Hollister, L. E., Motzenbecker, F. F., and Degan, R. O. (1961). Withdrawal reactions for chlordiazepoxide (Librium). *Psychopharmacologia* **2**, 63–8.

International Classification of Diseases (1977). Manual of the international statistical classification of diseases, injuries and causes of death (ninth revision) WHO, Geneva.

Iversen, L. L. (1978). Biochemical pharmacology of GABA. In *Psychopharmacology, a generation of progress* (eds. D. Lipton, A. Mimascio, and K. Killam), pp. 25–38. Raven Press, New York.

Kales, A., Scharf, M. B., and Kales, J. D. (1978). Rebound insomnia: a new clinical syndrome. *Science* **201**, 1039–41.

——, ——, ——, and Soldatos, C. R. (1979). Rebound insomnia: a potential hazard following withdrawal of certain benzodiazepines. *J. Amer. Med. Assoc.* **241**, 1692–5.

Klein, H. E., Broucek, B., and Greil, W. (1981). Lithium withdrawal triggers psychotic states. *Brit. J. Psychiatry* **139**, 255–6.

Kramer, J. C., Klein, D. F., and Fink, M. (1961). Withdrawal symptoms following discontinuation of imipramine therapy. *Amer. J. Psychiatry* **118**, 549–50.

Lader, M. (1983). Benzodiazepine withdrawal states. In *Benzodiazepines divided. A multi-disciplinary review* (ed. M. R. Trimble), pp. 17–32. John Wiley and Sons, Chichester.

——, Baker, W. J., and Curry, S. (1980). Physiological and psychological effects of clorazepate in man. *Brit. J. Clin. Pharmacol.* **9**, 83–90.

Marks, J. (1978). *The benzodiazepines: use, overuse, misuse, abuse*. MTP Press, Lancaster.

Millan, M. J. and Duka, T. H. (1981). Anxiolytic properties of opiates and endogenous opioid peptides, and their relationship to the action of benzodiazepines. In *Modern problems of pharmacopsychiatry* (eds. T. A. Ban, P. Pichot, and L. W. Pölvinger), Vol. 17, pp. 123–41. Karger, Basel.

Möhler, H. and Okada, T. (1977). Benzodiazepine receptors demonstrated in the central nervous system. *Science* **198**, 849–51.

Murphy, S. M., Owen, R. T., and Tyrer, P. J. (1984). Withdrawal symptoms after six weeks treatment with diazepam. *Lancet* **ii**, 1389.

Oswald, I. (1986). Drugs for poor sleepers? *Brit. Med. J.* **292**, 715.

Owen, R. T. and Tyrer, P. (1983). Benzodiazepine dependence. A review of the evidence. *Drugs* **25**, 385–98.

Petursson, H. and Lader, M. (1981). Benzodiazepine dependence. *Brit. J. Addiction* **76**, 133–45.

Power, K. G., Jerrom, D. W. A., Simpson, R. J., and Mitchell, M. (1985). Controlled study of withdrawal symptoms and rebound anxiety after six week course of diazepam for generalised anxiety. *Brit. Med. J.* **290**, 1246–8.

Preskorn, H. and Denner, J. (1977). Benzodiazepines and withdrawal psychosis. Report of three cases. *J. Amer. Med. Assoc.* **237**, 36–8.

Rickels, K., Case, W. G., Schweizer, E. E., Swenson, C., and Fridman, R. B. (1986). Low dose dependence in chronic benzodiazepine users. A preliminary report on 199 patients. *Psychopharmacology Bull.* **22**, 407–15.

——, ——, Winokur, A., and Swenson, C. (1984). Long-term benzodiazepine therapy: benefits and risks. *Psychopharmacology Bull.* **20**, 608–15.

Rosenberg, H. C. and Chiu, T. H. (1985). Time course for development of benzodiazepine tolerance and physical dependence. *Neuroscience and Biobehavioural Review* **9**, 123–31.

Russell, M. A. H. (1976*a*). Tobacco smoking and nicotine dependence. In *Research advances in alcohol and drug dependence* (eds. R. J. Gibbens *et al.*), Vol. 3. Wiley, New York.

—— (1976*b*). What is dependence? In *Drugs and drug dependence* (eds. M. A. H. Russell, D. Hawks, and M. MacCafferty) pp. 182–87. Saxon House/Lexington Books, Westmead.

Rutherford, D. M., Okoko, A., and Tyrer, P. J. (1978). Plasma concentrations of diazepam and desmethyldiazepam during chronic diazepam therapy. *Brit. J. Clin. Pharmacol.* **6**, 69–73.

Shur, E., Petursson, H., Checkley, S., and Lader, M. (1983). Long-term benzodiazepine administration blunts growth hormone response to diazepam. *Archives of General Psychiatry* **40**, 1105–8.

Skegg, D. C. G., Doll, R., and Perry, J. (1977). Use of medicines in general practice. *Brit. Med. J.* **2**, 1561–3.

Snaith, R. P. (1984). Benzodiazepines on trial. *Brit. Med. J.* **288**, 1379.

Squires, R. F. and Braestrup, C. (1977). Benzodiazepine receptors in rat brain. *Nature* **266**, 732–4.

——, Benson, D. I., Braestrup, C., Coupet, J., Klepner, C. A., Myers, J., and Beer, B. (1979). Some properties of brain specific benzodiazepine receptors: new evidence for multiple receptors. *Pharmacology, Biochemistry and Behaviour* **10**, 825–30.

Tyrer, P. and Seivewright, N. (1984). Identification and management of benzodiazepine dependence. *Postgraduate Med. J.* **60**, (supplement 2), 41–6.

——, Rutherford, D., and Huggett, T. (1981). Benzodiazepine withdrawal symptoms and propranolol. *Lancet* **i**, 520–2.

——, Owen, R., and Dawling, S. (1983). Gradual withdrawal of diazepam after long-term therapy. *Lancet* **i**, 1402–6.

Vessel, E. S. and Penno, M. B. (1983). Assessment of methods to identify sources of interindividual pharmacokinetic variations. *Clin. Pharmacokinetics* **8**, 379–409.

Winokur, A., Rickels, K., Greenblatt, D. J., Snyder, P. J., and Schatz, N. J. (1980). Withdrawal reaction from long-term low dosage administration of diazepam. *Archives of General Psychiatry* **37**, 101–5.

Winstead, D. K., Anderson, A., Eilers, M. K., Blackwell, B., and Zaremba, A. L. (1974). Diazepam on demand. Drug-seeking behaviour in psychiatric inpatients. *Archives of General Psychiatry* **30**, 349–51.

Woolf, J. H. and Nixon, J. C. (1981). Endogenous effector of the benzodiazepine binding site: purification and characterisation. *Biochemistry* **20**, 4263–9.

Yanagita, T. (1982). Dependence producing effects of anxiolytics. In *Psychotropic agents,* (eds. F. Hoffmeister and G. Stille), Part 2, pp. 395–406. Springer-Verlag, Berlin.

11

Psychopharmacology and social psychology: complementary or contradictory?

JIM ORFORD

INTRODUCTION

The problem of what exactly constitutes an adequate science of addiction becomes more and more intriguing and puzzling: particularly problematic is the reconciling of two great traditions of relevant science, one the biological tradition, the other the tradition of social science. If the two traditions, or the positions that exponents of the two traditions would take towards each other, were to be parodied, it seems that social scientists would be likely to describe a lot of what has been discussed in this book as 'reductionist', that is taking behaviour out of context, ignoring the social context of behaviour, missing the socially-constructed meaning of behaviour, relying far too much on a drug-focused approach, and assuming that the functions served by drugs are those that can be assessed in the types of experiments described. The biologist, on the other hand, is likely to see social science as 'soft', uncontrolled, superficial, missing the point, failing to see that humans and animals have a certain biological disposition in common, and failing to see that there are things about addiction potential, for example, that can be best shown by studying animals in the kinds of controlled conditions that are possible with animals. Whereas, of course, for the social scientist work with animals may be completely irrelevant, because animals clearly do not share the same social context as humans.

The overall question is whether these two positions are reconcilable. It seems that holding the middle ground between them is increasingly difficult in an age of specialization. It is almost impossible for a social scientist to take other than a very amateur interest in relevant work in the biological sciences, and vice versa.

There are four separate aspects of addictive behaviour which an adequate science of addiction should take into account.

1. Firstly, a wide range of addictive behaviours or 'excessive appetites'

need to be accounted for in an adequate theory of addiction (Orford 1985).

2. Secondly, these behaviours can serve a wide range of functions. This book has dealt with a number of different functions that potentially addictive drugs may serve, but the picture can be complicated still further by the suggestion that, if the social context is taken into account, the functions which addictive drugs and behaviours can serve is potentially varied and vast.

3. The science of addiction has, on the whole, tended to look at the positive incentives for behaviour, and has tended to neglect the other side of the coin, namely the disincentives for behaviour. These may play an equally important part in explaining why some people become addictive about some things, and other people do not become addictive about the same things.

4. Addictive behaviour should be seen as a process rather than something that is static; the functions served by addictive behaviour may change during the addictive process.

THE RANGE OF ADDICTIVE BEHAVIOURS

What constitutes the phenomena that a Society for the Study of Addiction should be concerned about? Where are the boundaries around this particular subject? It has been difficult to draw those boundaries clearly; in fact, in the case of drugs, it has been extremely difficult.

In the 1920s, on the question whether or not tobacco was addictive, Sir Humphrey Rolleston was saying, 'That smoking produces a craving . . . is undoubted, but it can seldom be accurately described as over-powering, and the effects of its withdrawal . . . cannot be compared with the physical distress caused by withdrawal in morphine addicts. To regard tobacco as a drug of addiction is all very well in a humorous sense, but it is hardly accurate' (Rolleston 1926). Much later, in the 1960s, Seevers (1962), a Dent lecturer some years ago, was saying, 'By no stretch of the imagination can either nicotine or caffeine conform to any accepted definition of addiction'. The case of benzodiazepines is another interesting example; clear evidence has been presented in this book (Chapter 10) of the possibly-addictive potential of these drugs that, when they were first produced, were thought not to possess such potential.

In drawing the boundaries around this subject we should move considerably beyond drugs however. It could be argued, for example, that certain forms of eating behaviour fall clearly within them. What psychiatrists have been calling 'bulimia' for a long time, or 'binge eating', clearly bears a resemblance to compulsive types of drug-taking. The existence of behaviour therapies to control eating behaviour, and such treatments as by-pass surgery and jaw-wiring, attest to the prevalence of dissonance about eating behaviour, as does the range of commercial and self-help groups for controlling eating behaviour.

Sex is another interesting example. Professor Warburton (Chap. 3, Fig. 3.4) showed that sex appeared to uniquely combine relaxing and stimulating properties. Sex clearly bettered all the drugs discussed in this book, and therefore it is hardly surprising to find that sex can become addictive. The list of some of the terms that have been used to describe excessive sexual behaviour includes the Casanova type, compulsive promiscuity, compulsive sexuality, Don Juanism (or the Don Juan syndrome or complex), Don Juaniterism (the female variety), erotomania, hyperaesthesia, hypereroticism, hyperlibido, hypersensuality, hypersexuality, idiopathic sexual precocity, libertinism, the Messalina complex, nymphomania, over-sexuality, pansexual promiscuity, pathological multipartnerism, and so on. There are numerous historical and fictional examples of hypersexuality. *My secret life*, the eleven-volume Victorian autobiography of a man who seemed to devote most of his time to sexual conquest, is one fairly well-known example (Anonymous 1966). Samuel Pepys and James Boswell are just two British historical figures who have given evidence in their writings of being in a serious state of conflict about their sexuality. There seem to be clear examples of sex as addictive behaviour.

Finally, an extremely important example for a science of addictive behaviours, is gambling. Gambling is a particularly important example because the phenomenology of so-called 'compulsive gambling' is really very similar to some of the much more familiar types of addictive behaviour. The descriptions of loss of control over gambling behaviour which compulsive or addictive gamblers give are strikingly reminiscent of descriptions of losing control over drinking behaviour, for example.

The phenomenon of Gamblers Anonymous as a self-help organization is itself a dramatic illustration of the fact that gambling has to be considered to be an addictive behaviour. Here is a self-help group which has spread widely, (though not as widely as Alcoholics Anonymous), modelling itself upon Alcoholics Anonymous, having a very similar programme, a similar number of steps in the programme through which members have to progress, and indeed suggesting that anyone who seriously wants to overcome a gambling problem must abstain totally from it. In terms of their philosophy there is no possibility of return to normal gambling, in exactly the same way as Alcoholics Anonymous has held to the view over the years that for someone who is a real 'alcoholic' there is no possible return to controlled or social drinking.

Any serious science of addiction must take excessive gambling into account. This is a difficult challenge, but a particularly important one. There is some evidence from a recent community survey in Exeter (McCartney 1986) that the general public has accepted the idea that alcohol addiction may be a form of dependence, even a disease, and hence one with few moral overtones, but has certainly not accepted the same view with regard to excessive gambling, which is still seen as a moral weakness. Thus the general public are mostly not including gambling within the boundary of what they think of as addictive behaviour, but the professionals should.

THE RANGE OF FUNCTIONS THAT ADDICTIVE BEHAVIOURS CAN SERVE

The second aspect of addictive behaviour to be considered concerns the numerous functions which behaviours and drugs of addiction potential can serve for individuals. Professor Warburton cited (Chap. 3) nicotine as a good example of a drug that can, in fact, serve both relaxation and stimulating functions. His theory was a functional one, that different people were using nicotine for different purposes, and indeed that the same people were using it for different purposes at different times. It is also evident that drugs such as the opiates are also complicated. They can produce apparently paradoxical, even apparently opposite, effects depending on dose, circumstances, and time since administration. So, even at this level of analysis, it is clear that these substances are really quite complicated, that they can work different kinds of magic for different people, and they can work different kinds of magic at different times.

What about gambling? In the development of a science of addiction, the functions of gambling need to be understood. So far, relatively little progress has been made in this field. There have been some demonstrations that heart rate increases when people gamble. For example, Leary and Dickerson (1985) showed that the heart rates of regular poker-machine players were raised by up to 20–30 beats a minute when they were playing the machines. A group from New South ˙ Wales (Blaszczynski *et al.* 1984) have looked at horse-race gamblers, who have shown not quite such large, but nevertheless definite, heart-rate increases when they placed bets.

Other people have looked at the functions that gambling might be serving in other ways. For example, Jacobs (1986) has looked at four components of what he calls 'dissociative states' associated with gambling. People, including regular gamblers were asked about their preferred activity, and other activities. One question was, 'Did you ever feel like you had been in a trance when undertaking the following activities?' About 80 per cent of regular gamblers said that they had felt that they had been in a trance when gambling. About 60 per cent of regular, heavy drinkers said the same, but gambling was the addictive behaviour which achieved the higher rate. Another question was: 'Did you ever feel like you had taken on another identity?' Again, 80 per cent of gamblers said they had, over 70 per cent of drinkers said they had, and a number of over-eaters said they had as well, but again gambling was the highest. 'Did you ever feel you were outside yourself, watching yourself do these activities?'—50 per cent of gamblers said that they had. Finally, 'Did you ever experience a memory black-out for a period when you had been doing this particular activity?' Not surprisingly, drinkers scored highest on that question; over 70 per cent of excessive drinkers said they had had memory black-outs, but so did 40 per cent of regular gamblers. Jacobs' argument here is that the function that gambling is serving is the function of blurring reality-testing, lowering self-criticalness and self-consciousness, and permitting what he called 'complimentary day-dreams' about the self. Which is very reminiscent of Hull's (1981) self-awareness theory of the function of alcohol. Hull looked at the evidence for the tension-reducing effects of alcohol and, like others such as Cappell and Herman (1972) before him, found that the evidence for the tension-reducing effect of alcohol was really quite equivocal. From the evidence that he looked at, Hull came to the view that alcohol, for many people who were abusing it, was in fact serving this reduced self-awareness function of cutting the drinker off from objective feedback, perhaps uncomplimentary and unpleasant, about the self. Now this very similar view is coming forward from Jacobs in the context of excessive gambling.

Another approach to the functions of gambling has been taken by Anderson and Brown (1986) in Glasgow, employing reversal theory (Apter, 1982). This is the theory whereby what is considered by a person to be an optimal state of arousal depends very much on the circumstances, and the pre-existing state, of the individual. In particular, the theory talks about telic and paratelic states. The telic state is future-oriented, involving planning, concentrating on foci outside self, enjoying the pleasure of goal anticipation, being meaningfully employed, and preferring to remain in a state of low intensity and low arousal. This is the kind of state found when people are working. Paratelic states, on the other hand, are said to exist when people are present-oriented, spontaneous, sufficient unto themselves, enjoying the pleasures of immediate sensations, and preferring states of high intensity and high arousal. This is the kind of state a person is in when playing.

The theory is that what is experienced as unpleasant arousal in a telic state might be experienced as excitement in a paretelic state. So the optimum level of arousal is very different; what in a telic state might be experienced as steady and alert might in a paratelic state be felt as boredom. The theory then goes on to say that this is a matter of individual differences, (some people may chronically prefer a paratelic to a telic state) but that it is also a question of time of day, and circumstances and aspects of the context; these factors will strongly affect a person's perception of the value of a drug or a potentially addictive behaviour. Another example is that the effect of alcohol at lunch-time is totally different from the effect of alcohol in the evening for many people. This may be because, brought up in the puritan ethic, those people like to maintain themselves in a telic state during working hours, and allow themselves to enter a more paratelic state in the evening.

However, complicated though this is, and a challenge though it may be to describe the personal functions of gambling and of other potentially addictive behaviours and substances, the situation gets much, much more complicated when social situations and self-attributed motives, functions, and meanings for addictive behaviours are considered. For example, the social psychologist, Sadava (1975), described the different functions that can be served by the same drug depending upon whether the user is a young person living in an urban slum, or a 'physician addict'. From the literature, he produced the following list of possible functions that might be served by opiates among young slum dwellers; to achieve detachment, to reduce threats to feelings of adequacy, to reduce drive states, to suppress pain and discomfort, to achieve a state of aloofness and isolation, to forget personal problems, to cope with sexuality, to help control social anxiety, to protect against depression, psychotic reaction or sadistic impulses, etc. For the physician addicts, on the other hand, the same drug might provide an alternative to excess alcohol use and relieve alcohol hangovers, relieve role strain inherent in the job, relieve the conflict between the activity of their job and a basic passivity, enhance feelings of omnipotence, etc. Thus the same drug is likely, he argues, to be serving very different functions depending upon the age, stage, environment, and role position of the person using the drug.

There are also likely to be important gender differences. In some work done in Exeter with men and women with alcohol problems consulting a clinical service, some significant differences between the functions attributed by men and women to their alcohol use were discovered (Orford and Keddie 1985). Briefly, the men were much more likely to attribute positive functions to moderate drinking than were the women.

For example, the men were much more likely to say that, although alcohol might be harmful in other ways, nevertheless in moderate doses it 'made them cheerful', it 'exaggerated whatever mood they were in', it 'made them happy', it 'made them feel fine', 'happy-go-lucky', it 'helped them meet new people', they 'liked going to the pub and playing pool', it 'made them more sociable', it was 'for companionship purposes', etc. Women were much less likely to mention those positive social functions of drinking, although men and women were equally likely to mention coping functions such as drinking enabling them to feel 'quieter', 'less frightened', 'less worried', 'more relaxed' and 'confident'. This highlights an important gender difference in functions served by the same drug.

Historically, there are likely to be differences. One can only imagine that functions being served by opium for the Victorian opium users described by Virginia Berridge (1977) in the Fenland district of East Anglia, or the Lancashire cotton mills, were very different from the kinds of functions being served by opiates for modern urban dwellers or modern physician addicts.

Then if some of the functions that may be served within social systems are considered, the situation becomes even more complicated. Some of the work on families is relevant here, as are the suggestions that have been made and studied by people like Steinglass (1982) on the functions that may be served for family systems, for example by excessive drinking. Some recent work by Jacob (1986) looks at families, and at the functions that may be served by drinking and the very different significance that excessive drinking may have, depending on the pattern and style of the drinking. He has drawn particular attention to the distinction between two types of drinking behaviour in families, both apparently constituting a problem. One he calls 'binge out-of-home drinking', where most of the drinking takes place away from home and in a binge-like pattern, and he contrasts that pattern with what he calls 'steady in-home drinking', where the drinking pattern is much steadier, and where most drinking takes place in the home.

Looking at interaction in a controlled setting between husbands and wives (the husbands being the ones with the identified drinking problems) Jacob found that interaction was much more negative in those families where the excessive drinking was 'binge out-of-home'. In the 'steady in-home' excessive drinkers, interaction was no more negative than in a control group. Furthermore, when he looked at what happened over a period of days, depending on whether there had been excessive drinking, in the 'binge out-of-home' drinkers, a period of drinking was followed by a decrease in satisfaction on the part of the wives, but this was not the case in the 'steady in-home' drinkers. The findings are complex, but this certainly begins to look as if the same drug being taken within families, within the same geographical and historical context, may be serving very different functions for the family as a social system.

NEGLECT OF THE DISINCENTIVES

Most of the science of addiction so far has looked in the direction of positive incentives for drug use, or positive reasons why people abuse drugs, and has tended to neglect the possible reasons why those who do not abuse drugs or other potentially addictive agents do not do so. It is as if we have looked at one half of the explanation, but not the other.

The log-normal or very skewed distribution curve for alcohol consumption in a population is familiar. It turns out that there are similarly skewed, approximately log-normal distribution curves for other potentially addictive behaviours. Kinsey and his colleagues (1948), for example, in their classic book on the sexual behaviour of the American male, had a graph that looks remarkably similar, and which depicts number of orgasms per week. It has a similar shape, with the majority of men being relatively moderate in frequency, and with decreasing numbers of people being ever more extreme and having ever greater numbers of orgasms per week. It is difficult to find gambling data of a similar kind, but, from an American survey by Kallick and colleagues (1979), exactly the same thing was illustrated, with the majority of people being abstinent or very moderate in gambling, but with a marked skew to the frequency distribution curve. The same can be shown with drugs (Orford 1985, p. 163).

One possible theoretical explanation is that these curves are 'conformity curves', and here we have to go back to the social psychology of the 1930s and Allport's (1934) description of these curves, which he called 'J-shaped' curves. His understanding of such curves was that they represented the imposition of social conformity upon people's natural inclinations to behave in various, and sometimes irresponsible, ways. If you look, for example, at the speed at which people drive across road junctions where they are supposed to halt, you find that the majority of people, as you would expect, either halt or slow down. But some people slow down less than others, and the distribution curve for the speed at which people cross such junctions shows this familiar skewed frequency distribution form. The argument of social psychologists, such as Allport, at that time was that this represented social conformity imposed upon the natural but variable tendency to go at speed across the junction. In the present discussion, the argument would be that these curves, familiar to us now for describing the frequency distribution of appetitive or potentially addictive behaviours, are in fact due to conformity; thus one whole side of the behaviour equation has been missed, the side which has to do with people's tendencies to conform to abstinence or moderation norms.

The disincentives are also likely to include some of the pharmacological consequences of using a drug that may be aversive and which may therefore detract from people's inclinations to use the substance, as well as people's involvements in other activities, which simply preclude them from being heavy drug users, because they are just too busy working, being educated, and getting on with their family lives. The argument is that many people who do not use opiates or other drugs or who do not use them much, might find them just as functionally useful as people who do; it is simply that they are too conformist or too busy to try. The disincentives are just too great.

It is interesting to compare this field with one like criminology, which has been more influenced by the social sciences, and where the idea of deterrence, or the idea of controls on criminal behaviour, is much better understood and much more widely recognized. More credence is given to the view that many more of us would be delinquents if it were not for the fact that we have some perception of the probability of being caught, and of receiving some of the penalties that we expect would follow. Or alternatively that we are too much engaged in other activities to bother with delinquent behaviour, or just too well socialized towards conformity.

ADDICTION AS A DYNAMIC PROCESS

Addictive behaviour is a process that changes through time. A great deal of the science of addiction has been in a fairly static mould, assuming that we can predict liability to dependence on the basis of the fixed properties of drugs, or of individual people. However, if much of the evidence from the social sciences in this field were taken seriously, it is apparent that addiction is a changing process.

Examples can be found in work by people like Robins and her colleagues (1977) on Vietnam veterans, and the slightly less well-known work by Kandel (1978) on the process of becoming a serious drug-taker amongst young people in New York State. Both of these works, and a number of others reported since, show that the factors that predict transition from one stage in the process to the next may be very different, depending upon the stage that has been reached. To cut a long story short, the suggestion is that some of the intrinsic factors of interest, that is some of the euphoriant effects, some of the relaxing effects, some effects that animal studies can show us much more clearly than human studies, become more important at a relatively advanced stage in a drug-using career. The early stages of drug-taking are more likely, in general, to be under extrinsic social control. The factors that influence early drug-taking, or the uptake of potentially addictive behaviour at an early stage, are much more likely to be social factors (for example what a peer group is doing, or the pressures that are put upon a person socially) than they are to be the intrinsic rewards of that behaviour.

Some early results of a study at Exeter, looking at the outcome for a sample of young adults between the ages of 16 and 35 who have had parents with drinking problems, are relevant here. There is considerable evidence that, to an extent, alcohol and other drug problems run in families, and the study was to look at some of the reasons why that might be. For example, when the drug-taking of these young adults, and a control group is compared, it transpires that illicit drug-taking *per se* is no more frequent amongst those who have had parents with drinking problems. What is different is the level of involvement in drug-taking. In other words, when more 'advanced' drug-taking is considered (for example, regular drug-taking, or drug-taking that produces problems) then those young adults who had parents with drinking problems begin to appear more frequently than do those in the comparison sample.

CONCLUSION

Lest the field of psychopharmacology be open to accusations of reductionism and examining addictive behaviour out of context, work in the psychopharmacology of addiction must take account of principles and findings derived from social science studies in the same area.

A science of addiction must be capable of accounting for a wide range of substances and activities which have addiction potential. These include a wider range of drugs than was once thought, as well as a number of non-drug activities, of which excessive gambling provides perhaps the greatest challenge to theories of addiction.

Addictive behaviours are capable of serving a very broad range of personal and social functions, not all of which can be properly represented in controlled, laboratory conditions. The particular functions served can depend upon gender, social status, ethnic group, culture, and historical epoch, as well as upon dose, time since ingestion or activity, setting, and expectation.

A full account of addictive behaviour may need to give greater prominence to the deterrence side of the attraction-deterrence equation which determines a person's particular appetitive behaviour. The disincentives or restraints that control behaviour may be as influential as the incentives or attractions. The former have been relatively neglected in previous theories of addiction.

It may be more productive to view addiction as a dynamic process rather than as a static entity. Progress has been made in understanding drug-taking careers, for example, by breaking the process down into a series of stages. The factors that predict transition from one stage to another may be very different at 'early' and 'late' stages.

These points cover four of the areas that an adequate science of addiction must take into account. It seems appropriate to finish this chapter with a quotation from a learning theorist of the early 1970s who is taking what looks like a typical reductionist view (Dews 1973, p. 45). He wrote, with reference to learning schedules:

'Psychic dependence is nothing more or less than schedule control of drug-taking behaviour.... Although the mechanisms are normal, they operate in a mechanistic way, so the consequences may be detrimental to the subject, according with the well recognized "irrationality" of drug-taking ... It has been the search for the rationality of the drug-takers' conduct that has inhibited suggestions of behavioural mechanisms ... The addict's account of how he came to his condition should not be taken any more seriously than a cancer victim's account of how he came to be so afflicted; in both cases, processes are involved into which the subject has no particular insight.'

And to a social scientist, of course, that statement would be like a red rag to a bull.

REFERENCES

Allport, F. (1934). The J-curve hypothesis of conforming behaviour. *J. Social Psychol.* **5**, 141–81.
Anderson, G. and Brown, R. I. F. (1986, unpublished). 'Some applications of reversal theory to the explanation of gambling and gambling addictions.' Department of Psychology, University of Glasgow.
Anonymous (1966). *My secret life*. Grove Press, New York.
Apter, M. J. (1982). *The experience of motivation: The theory of psychological reversals*. Academic Press, London.
Berridge, V. (1977). Fenland opium eating in the nineteenth century. *Brit. J. Addiction* **72**, 275–84.
Blaszczynski, A. P., Winter, S. W., and McConaghy, N. (1984). 'Plasma endorphin levels in pathological gambling.' Paper presented at the Sixth National Conference on Gambling and Risk Taking, Atlantic City, New Jersey. December 9–12.
Cappell, H. and Herman, C. (1972). Alcohol and tension reduction: a review. *Quart. J. Studies on Alcohol* **33**, 33–64.

Dews, P. (1973). The behavioral context of addiction. In *Psychic dependence: Definition, assessment, in animals and man, theoretical and clinical implications* (eds. L. Goldberg and F. Hoffmeister). Springer-Verlag, Heidelberg.

Hull, J. (1981). A self-awareness model of the causes and effects of alcohol consumption. *J. Abnormal Psychology* **90**, 586–600.

Jacob, T. (undated, unpublished). 'Alcoholism and family interaction: clarifications resulting from subgroup analyses and multi-method assessments. Dept. of Psychology, University of Pittsburgh.

Jacobs, D. F. (1986, unpublished). 'Behavioral aspects of gambling: Evidence for a common dissociative-like reaction among addicts.' Loma Linda University Medical School, Loma Linda, California.

Jaffe, J. (1977). Tobacco use as a mental disorder: the rediscovery of a medical problem. In *Research on smoking behavior* (eds. M. Jarvik, J. Cullen, E. Gritz, T. Vogt, and L. West). National Institute on Drug Abuse Research Monograph 17, US Department of Health, Education and Welfare, NIDA, Rockville, Maryland.

Kallick, M., Suits, D., Dielman, T., and Hybels, J. (1979). A survey of American gambling attitudes and behavior. University of Michigan, Survey Research Centre, Institute for Social Research.

Kandel, D. (ed.) (1978). *Longitudinal research on drug use: Empirical findings and methodological issues.* Hemisphere, Washington.

Kinsey, A., Pomeroy, W., and Martin, C. (1948). *Sexual behavior in the human male.* Saunders, Philadelphia.

Leary, K. and Dickerson, M. (1985). Levels of arousal in high- and low-frequency gamblers. *Behaviour Research and Therapy* **23**, 635–40.

McCartney, J. (1986, unpublished). 'Perceptions of addiction and change.' PhD. thesis, University of Exeter.

Orford, J. (1985). *Excessive appetites: A psychological view of addiction.* Wiley, Chichester.

—— and Keddie, A. (1985). Gender differences in the functions and effects of moderate and excessive drinking. *Brit. J. Clin. Psychol.* **24**, 265–79.

Robins, L., Davis, D., and Wish, E. (1977). Detecting predictors of rare events: Demographic, family and personal deviance as predictors of stages in the progression toward narcotic addiction. In *The origins and course of psychopathology: Methods of longitudinal research* (eds. S. Straus, H. Babigian, and M. Roff). Plenum Press, New York.

Rolleston, H. (1926). Medical aspects of tobacco. *Lancet* **i**, 961–5.

Sadava, S. (1975). Research approaches to illicit drugs: A critical review. *Genetic Psychology Monographs* **91**, 3–59.

Seevers, M. (1962). Medical perspectives on habituation and addiction. *J. Amer. Med. Assoc.* **181**, 92–8.

Steinglass, P. (1982). The roles of alcohol in family systems. In *Alcohol and the family* (eds. J. Orford and J. Harwin). Croom Helm, London.

Index

181